The Practice of Cost Segregation Analysis

Maximizing Tax Benefits for Building Acquistions and Construction

- Documenting & Preparing a Cost Segregation Study
- IRS Guidelines, Tax Regulations & Precedent-Setting Rulings
- Designing & Building for Optimum Tax Benefits

by Bruce A. Desrosiers, CPA, MST
and Wayne J. DelPico

RSMeans

WILEY

The Practice of Cost Segregation Analysis

Maximizing Tax Benefits for Building Acquistions and Construction

- Documenting & Preparing a Cost Segregation Study
- IRS Guidelines, Tax Regulations & Precedent-Setting Rulings
- Designing & Building for Optimum Tax Benefits

by Bruce A. Desrosiers, CPA, MST
and Wayne J. DelPico

RSMeans

WILEY

This book is printed on acid-free paper.

Published by John Wiley & Sons, Inc., Hoboken, New Jersey

Published simultaneously in Canada

The managing editor for this book was Mary Greene. The editor was Andrea St. Ours. The editorial assistant was Jessica Deady. The production manager was Michael Kokernak. The electronic publishing specialist was Paula Reale-Camelio. The proofreader was Robin Richardson. The book and cover were designed by Paula Reale-Camelio.

For general information on our other products and services, or technical support, please contact our Customer Care Department within the United States at 800-762-2974, outside the United States at 317-572-3993 or fax 317-572-4002.

Wiley also publishes its books in a variety of electronic formats. Some content that appears in print may not be available in electronic books. For more information about Wiley products, visit our web site at www.wiley.com.

Library of Congress Cataloging-in-Publication Data:

ISBN: 978-0-87629-744-5

10 9 8 7 6 5 4 3 2

Table of Contents

Part III Appendix, Glossary, & Resources

This book is dedicated to our wives and daughters for their love, support, and encouragement.

About the Authors & Reviewers

About the Authors

Bruce A. Desrosiers, CPA, MST, is a tax manager at Sullivan & Company, CPAs, LLP, of Providence, Rhode Island, where he is responsible for tax compliance, planning, and cost segregation services. Mr. Desrosiers joined Sullivan & Company as a tax specialist in 1997, having been previously associated with Ernst & Young. Sullivan & Company, one of the largest independent public accounting firms in southeastern New England, serves privately-held and family-owned businesses. The company provides full-service accounting expertise to diversified industries, including health care, real estate, construction, distribution and manufacturing facilities, and nonprofit and hospitality organizations.

Mr. Desrosiers holds a Bachelor of Science degree in Accounting from Providence College and a Master of Science degree in Taxation from Bryant College. He is a member of the Rhode Island Society of Certified Public Accountants.

Wayne J. DelPico is Vice President of the Mulligan Company, Inc., where he currently oversees residential and commercial development and building projects in Southeastern Massachusetts. He has more than 25 years' experience in construction management, and has been involved in projects throughout most of the United States.

Mr. DelPico holds a degree in Civil Engineering from Northeastern University in Boston, where he teaches construction estimating and project management. He is the author of *Estimating Building Costs for the Residential & Light Commercial Contractor* and *Plan Reading & Material Takeoff,* also published by RSMeans, and is a lecturer on construction estimating topics, including cost segregation, in seminar programs presented throughout the U.S.

About the Reviewers

Patrick J. Riegel is the Senior Manager of Cost Segregation Services for the Freed Maxick Group in Buffalo, New York. He has 20 years' experience in the construction industry, and over the past five years has completed more than 1,200 cost segregation studies ranging in size and complexity. Studies have included restaurants, manufacturing facilities, apartment complexes, office buildings, warehouses, malls, industrial plants, retail plazas, automobile dealerships, and high-rise office buildings.

Mr. Riegel is responsible for the detailed cost segregation of building components for new and existing construction projects, and supervises a team of estimators and appraisers at Cost Segregation Partners, the engineering division of the Freed Maxick Group. Prior to specializing in cost segregation, he was President and General Manager of a bonded general construction company with annual revenues of $7-$10 million. Mr. Riegel has Bachelor of Arts and Construction Technology degrees, as well as certifications in OSHA Construction Safety and Health Training.

Joseph Strickland, Jr., RA, author of the foreword, is Vice President-National Director of the Cost Segregation Group for American Appraisal Associates, where he leads a team of seven cost segregation professionals in Milwaukee and New York. Mr. Strickland has over 25 years of experience in architecture, construction, and valuation services, with particular emphasis on investment tax credit, cost segregation analysis, insurance claims analysis, project management, and allocation of purchase costs. His cost segregation background includes automobile assembly facilities, gaming and recreation facilities, restaurants, hotels, industrial and process facilities, financial institutions, retail establishments, office buildings, and steel plants, as well as cost segregation audit support at the IRS level.

Mr. Strickland has also served as a designer, project manager, owner, and owner's representative on single/multi-family residential, educational, industrial, commercial, retail, and land development projects. He holds a Master of Architecture and a Master of Education degree, and is a registered architect in Connecticut and Pennsylvania.

Foreword

In 1962, legislation was signed that provided an incentive for our nation's capital-intensive industries to rebuild or upgrade their physical plants. Through years of judicial and legislative interpretation, law-making, and major changes in 1986 and 1993, this incentive had been extended to federal taxpayers in all industries.

While those of us in the cost segregation profession rely on many court decisions to support our positions, the 1997 court case decision, *The Hospital Corporation of America v. Commissioner*, is particularly significant in that it further expanded the possibilities for accelerating the timing of depreciation for certain assets properly classified, or "segregated," using rules developed during the ITC era. Cost segregation studies have become an important element in tax planning for all federal taxpayers who are depreciating assets.

The new building on your balance sheet—or the property you just acquired or renovated—may generate more cash flow benefits than you realize. The costs of short-lived assets contained in construction or acquisition documentation may not be readily apparent. A cost segregation study—a service that exposes these costs—can be a potent tool, bringing to light substantial tax savings made possible by accelerated depreciation.

Building owners and tenants have an opportunity to recognize depreciation benefits under current tax law if they are:

- Building a new facility
- Renovating, remodeling, or expanding an existing building
- Purchasing or acquiring real estate
- Relocating and investing in leasehold improvements

Proper depreciation of assets continues to present a significant opportunity to reduce your current federal income tax liability. Without cost segregation or purchase cost allocation, however, many property owners fail to fully capture these benefits. A cost segregation analysis breaks down construction and acquisition costs and allocates them to specific categories of tangible personal property, land improvements, and real property.

The results of a cost segregation study, depending on effective tax rates and discount rates, can provide substantial savings on a net present value basis. These savings are realized by the accelerated depreciation of property when it is reclassified from a long life to shorter life depreciation categories.

Without a cost segregation study, it is difficult for a property owner to separate all of the personal property and land improvement costs, as well as indirect costs, from the total cost of the building. Without segregating these costs, all such property is generally subject to a 39-year straight-line depreciation (27.5 years for residential real property)—resulting in a substantially higher current tax liability.

The success of a cost segregation study lies in a thorough understanding of tax laws and construction documentation. This book provides a foundation upon which to build cost segregation knowledge in both of these areas. A credible cost segregation study requires a detailed audit trail that can only be created by applying some of the techniques discussed in this book. This publication is a valuable resource among the references needed by both practicing cost segregation professionals and those who wish to have a better understanding of how these studies are created.

— *Joseph Strickland, Jr., RA*
Vice President-National Director of the Cost Segregation Group for American Appraisal Associates

Chapter 1 Introduction to Cost Segregation

"In this world, nothing can be said to be certain except death and taxes."
– Benjamin Franklin (1706-1790)

While both death and taxes may be inevitable, we are constantly challenged to figure out a way to defer both. The potential value of deferring taxes is one of the primary reasons there is such a growing interest in cost segregation.

What Is Cost Segregation?

Cost segregation is the practice of identifying assets and their costs, and classifying those assets for federal tax purposes. Sometimes called "accelerated cost recovery," cost segregation analysis is an increasingly popular phenomenon, as facility owners understand the opportunity it offers to defer taxes, reduce their overall current tax burden, and free up capital by improving their current cash flow. Cost segregation studies can result in accelerated depreciation deductions for properties including new buildings being constructed, renovations of existing buildings, leasehold improvements, and the purchase of real estate.

Evolution and Benefits

Opportunities for cost segregation grew out of the court case *Hospital Corporation of America v. Commissioner* in 1997. The resulting rulings enabled practitioners to support significant tax benefits related to depreciation of property. The Tax Court ruled that the Investment Tax Credit rules should be used by the taxpayer to make asset classifications. Certain costs previously classified as subject to a 39-year depreciable life could be classified instead as personal property or land improvements, with a 5-, 7-, or 15-year rate of depreciation using accelerated methods.

To understand the significance of this change, for every $100,000 of investment that can be converted from a 39-year to a 7-year depreciable life, the result is a $20,000 improvement in cash flow (present value of the the tax benefits assuming the highest marginal tax rate, using a discount rate of 8% over the life of the building). Taxpayers can reclassify assets that had been depreciated less favorably to personal property classifications—going back as far as 1987. Most structures placed in service prior to 1987 were depreciated over much shorter lives, and like personal property, would be almost fully depreciated by now, rendering cost segregation of these facilities less favorable.

The 2003 Jobs and Growth Tax Relief Reconciliation Act (JGTRRA) expanded the bonus depreciation from 30% to 50%—for qualifying assets acquired after May 5, 2003. This allows a 50% deduction of the cost of new MACRS (Modified Accelerated Cost Recovery System) property if it has a class life of up to 20 years. This means that an investment of $100,000 in land improvements can yield a $50,000 immediate deduction, with the rest of the cost deductable over a normal recovery period. *(See Chapter 2, "Income Taxes & Depreciation Deductions," for more on bonus depreciation.)*

Candidates for Cost Segregation

Among the properties likely to benefit from a cost segregation study are those with a major investment in site work (e.g., paving, curbing, underground utilities, drainage, signage, and fencing). Site work, considered land improvements, has a depreciable life of 15 years, as compared to a building's traditional 39-year life. Interior components, such as movable partitions and business-related equipment and its associated piping or wiring, are some examples of assets that can be classified as personal property with a 5- or 7-year depreciable life.

Restaurants, supermarkets, hospitals, and manufacturing facilities are types of buildings that especially benefit from cost segregation studies. Highly specialized facilities can result in a majority of the assets being reclassified to classifications that bring about faster recovery through depreciation deductions.

Who Performs Cost Segregation?

Cost segregation studies are a specialty offered by accounting firms and consultants. Developing cost segregation studies requires not only tax and accounting knowledge, but the expertise of construction/engineering and cost estimating professionals. In some cases, accounting firms may work together with outside engineering professionals. There are also many consultants who specialize in all the elements of cost segregation.

To oversee a cost segregation study, accounting/tax professionals must have a background in classification issues relating to real estate construction and depreciation. The construction/engineering member of the cost segregation study team brings knowledge of construction, including plan and specifications reading, quantity takeoff, and estimating procedures. The tax/accounting professional advises the construction expert on which building components qualify for classification as "other than building components," and therefore accelerated depreciation.

The accountant classifies building components and tangible personal property as appropriate, then calculates the depreciation. When a change to prior year classifications is made, the accountant will recalculate the depreciation, and the difference in the methods is taken as a deduction in the current year. (Taxpayers may be required to file IRS Form 3115 to gain consent or approval for the change, as explained in Chapter 2.)

What Are the Steps in a Cost Segregation Study?

Cost segregation studies typically require the following steps:

- Feasibility/applicability: Evaluation of the building owner's current tax status and future business plans to determine the value/appropriateness of performing a cost segregation analysis.
- Evaluation of the building's or facility's construction costs, by component or system comprised of components.
- A detailed review of the project's construction documents, including plans (architect's and engineer's drawings) and specifications.
- A visit to the facility to identify building components and systems, and how they are used.
- A detailed "engineering" review of building systems, including walls, flooring, ceilings, telecommunications, plumbing, mechanical, and electrical. (Note: Other design and construction professionals, besides engineers, are also qualified to perform this type of review. The key is an understanding of building systems and the ability to generate credible takeoffs and estimates.)
- Classification (or reclassification, in an existing building) of building systems and components for accelerated depreciation.
- Identification and allocation of indirect project costs.
- Compilation of reports and completion of tax forms as needed to support asset classifications, calculations, and deductions.

Specific Requirements

The IRS's *Cost Segregation Audit Techniques Guide*, made available to the public in 2004 and available online (at http://www.irs.gov/pub/irs-utl/costsegatg.htm), advises field auditors that its view of a "quality cost segregation study" and its principal elements include:

- Preparation by an individual with expertise and experience
- Detailed description of the methodology
- Use of appropriate documentation

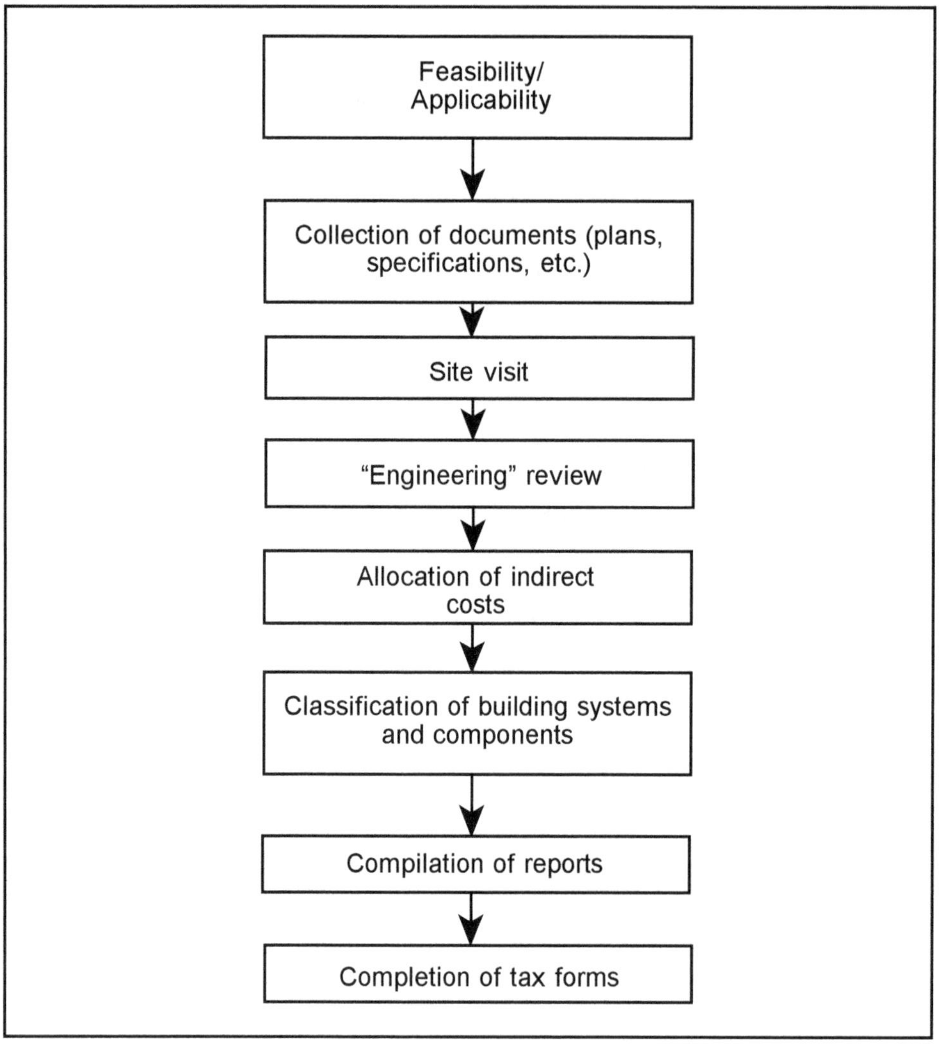

Figure 1.1

- Interviews conducted with appropriate parties
- Use of a common nomenclature
- Use of a standard numbering system
- Explanation of the legal analysis
- Determination of unit costs and engineering "takeoffs"
- Organization of assets into lists or groups
- Reconciliation of total allocated costs to total actual costs
- Explanation of the treatment of indirect costs
- Identification and listing of Section 1245 property
- Consideration of related aspects (e.g., IRC § 263A, Change in Accounting Method and Sampling Techniques)

Note: The IRS's Cost Segregation Audit Techniques Guide *is not to be used as a means of documenting the IRS position and may not be cited as a precedent. It is an internal publication for IRS field auditors, which can provide some insight to readers. Detailed rulings and numerous other documents are published on the IRS Web site, which should be checked periodically for updates. (This book's Web site, http://www.rsmeans.com/supplement/67345.asp, will make note of updates when they become available.)*

The IRS's *Cost Segregation Audit Techniques Guide* also suggests that a quality cost segregation report include these elements:

- A summary letter/executive summary
- Narrative report
- Schedule of assets
- Schedule of direct and indirect costs
- Schedule of property units and costs
- Engineering procedures
- Statement of assumptions and limiting conditions
- Certifications
- Exhibits

(See Appendix A of this book for further explanation of the cost segregation study "principle elements" and elements of the cost segregation report from the IRS's Cost Segregation Audit Techniques Guide.)

How This Book Supports Cost Segregation Study Requirements

The chapters that follow offer guidance to support the development of a cost segregation study. Part I, "Tax Regulations & Precedent-Setting Rulings," focuses on IRS cases, court rulings, and classification and depreciation methods. Chapter 2 begins this coverage by reviewing taxable income regulations and deductions, including those for depreciation. Chapter 3 identifies important source materials, such as IRS documents, that can be used to clarify tax positions. Chapter 4 reviews important court cases with the rulings and guidance from the IRS relating to depreciation. Proper classification of property is a central concept in cost segregation; Chapter 5 reviews the different types of classifications and how they are determined.

Part II, "The Engineering/Estimation Portion of Cost Segregation," outlines the cost estimating tasks involved in a cost segregation study. These begin in Chapter 6, with a review of the methodologies recognized by the IRS for asset allocation. This chapter points out the particular requirements, the needed expertise, and the cautions associated with each method, along with guidelines for the site visit. Chapter 7 explains construction contract documents, including the plans and specifications that must be examined as part of the cost segregation study—both to determine quantities and costs and to evaluate how building systems are used for depreciation purposes.

Chapter 8 explains construction estimating and quantity takeoff procedures that the cost segregation team members performing the estimating portion of the study need to be familiar with in order to assemble cost estimates to validate the costs of assets identified for shorter depreciable lives. Chapter 9 is a review of area and volume calculations used by construction estimators. This information is useful for cost segregation analysts in "taking off" quantities as part of the estimating process.

Chapter 10 explains how construction professionals assemble material and labor costs, so that the cost segregation analyst can create an audit trail for all of the costs associated with an asset. Chapter 11 provides information on reliable construction cost data sources that are accepted or recognized by construction industry professionals and the Internal Revenue Service.

New construction projects offer the opportunity to optimize cost segregation. Project documentation can be organized, and building systems selected in ways that strongly support the benefits of cost segregation—from the start. Chapter 12 addresses the actions each "team member" (architect/engineer, contractor, and cost segregation analyst) should take at the early planning stages of a new construction project. The chapter also notes the opportunities that exist for design and construction professionals who wish to expand the services they offer to building owners—from guidance in the earliest stages of the project, to delivery of documentation and invoices—in a format that will facilitate a cost segregation study.

IRS Updates

While there have not been significant changes to the depreciable lives of non-residential buildings since the 1993 Tax Act (which extended the life of these buildings to 39 years—up from 31.5 years), the American Jobs Creation Act of 2004 (enacted October 22, 2004), introduced some new classifications for non-residential buildings. These include "Qualified Restaurant Property" and "Qualified Leasehold Improvement Property." This act treats the new classifications as 15-year property. *(See Chapter 5 for more on the effects of the act.)*

It is important to be aware of updates in IRS and federal government regulations, rulings, and guidelines. Check the IRS Web site or this book's Web site (http://www.rsmeans.com/supplement/67345.asp) for notification of updates.

Conclusion

Whether owners are acquiring an existing building; planning to construct, expand, or remodel one; or seeking to reduce taxes on facilities acquired or constructed since 1987, cost segregation offers a tremendous opportunity. By reducing their tax burden through accelerated depreciation of certain assets, they can increase their current cash flow.

Chapter 2

Income Taxes & Depreciation Deduction Methods

Before considering the potential benefits of a cost segregation study, it is helpful to have an understanding of the framework of income taxes and depreciation. The first part of this chapter is a review of types of income, as viewed by the Internal Revenue Service, and the way buildings and property are generally treated. The rest of the chapter focuses on the depreciation deduction and its impact on building owners' taxable income.

Understanding Income Tax

The Internal Revenue Code (IRC) imposes taxes on the taxable income of individuals, estates[1], trusts, and corporations. *Pass-through* entities are not taxed directly; operating results (income, deductions, and credits) are passed through to the shareholders, partners, or members, as the case may be. The items passed through retain their character until they are taxed to one of the above-mentioned taxpaying entities. The most common types of pass-through entities are partnerships and S-corporations.

The tax on individuals, estates, and trusts has ranged from a low of 10%, to a high of 39.6% over the past decade.[2] Recent tax legislation decreased the rates in the various income brackets.[3] An individual in the 39.6% bracket, for example, will pay that rate on each additional dollar of income (amount over that bracket) subject to tax. Stated differently, a decrease in income in a given year will result in savings for that year at the same rate. This is sometimes referred to as the "marginal tax rate." For the layman, this is the tax rate on the next dollar of income you earn. Example: Bob has taxable income of $20,000 and is in the 15% tax bracket. If he receives/earns an additional $1,000, he will pay an additional $150 of tax on that change in his income ($1,000 x 15%).

Taxable income is generally defined as gross income minus allowable deductions.[4] *Gross income* consists of all earnings from any source, including but not limited to the following:

1. Compensation for services including fees, commissions, fringe benefits, and similar items
2. Gross income derived from business
3. Gains derived from dealings in property
4. Interest
5. Rents
6. Royalties
7. Dividends
8. Alimony and separate maintenance payments
9. Annuities
10. Income from life insurance and endowment contracts
11. Pension
12. Income from the discharge of indebtedness
13. Distributive share of partnership gross income
14. Income in respect of a decedent
15. Income from an interest in an estate or trust[5]

Some of the above items may actually be excluded under certain circumstances.[6] Allowable deductions are subtracted from gross income items to arrive at taxable income. The courts have stated that the allowance of deductions is a matter of legislative grace. In other words, the deductions are allowable, but the burden of proving the validity of those deductions is with the taxpayer. The allowable deductions are specified by Internal Revenue Code §161 and its following sections. They include but are not limited to:

1. Trade or business expenses[7]
2. Interest[8]
3. Taxes[9]
4. Losses[10]
5. Bad debts[11]
6. Depreciation[12]
7. Amortization[13]
8. Expenditures to remove architectural barriers to the handicapped and the elderly[14]
9. Environmental remediation costs[15]
10. Other deductions[16]

The above expenses may not be allowed in certain circumstances.[17] Businesses that self-construct properties are required to capitalize direct and indirect costs associated with the construction,[18] including expenses that would otherwise be deductible under Sections 161-198 of the IRC.[19]

The cost of acquiring a new building or for permanent improvements to any property cannot be deducted in arriving at taxable income either.[20] This exception of ordinary and necessary business deductions is generally extended to any asset that can reasonably be expected to last more than one year. The costs capitalized under these circumstances are generally subject to the depreciation allowance, which is calculated under Section 168. The rest of this chapter will focus on depreciation deductions and the effect they have on taxable income.

Depreciation Deductions

Cost segregation professionals can have a direct impact on depreciation by classifying property to allow for deductions to be taken sooner in the life of constructed assets. This can be accomplished by a thorough analysis and classification of the costs of constructing a property. The increased deductions translate directly into lower taxable income in the earlier years of an asset's useful life. Consider the following example:

Big Building Co., a pass-through entity, performs services during the first year of operations and receives fees of $1,000,000. The company has operating expenses of $450,000, before depreciation deductions. Its depreciation deduction is $150,000. The resulting taxable income is $400,000. Taxable income is calculated by subtracting from the "gross income from fees" the operating expenses and depreciation ($1,000,000 - $450,000 - $150,000). If the owners of Big Building Co. have a 40% marginal tax rate, the tax on this activity would be $160,000 ($400,000 x 40%).

The company hires a cost segregation professional to perform a study, and the first-year depreciation is recalculated as $250,000. The taxable income is now $300,000 ($1,000,000 - $450,000 - $250,000), and the tax on that income is $120,000. The result is a decrease in the tax liability for the current year, a savings of $40,000. This savings may be reinvested in this business activity or other investments until later years of the asset's life when the savings reverse.[21]

This change in the timing of deductions should be communicated to the building owner as the net present value of tax savings and/or expense over the life of the asset. This is where the value of performing cost segregation studies becomes apparent. *Net present value* refers to the fact that having a dollar today is always better (more valuable) than waiting to get a dollar tomorrow or a year from now. Accountants usually measure the net present value using a rate of interest applied to cash flows over some time period.

There are some situations that may result in the increased depreciation having no value. In these cases, practitioners would be well-advised to know the taxpayer's specific circumstances relating to the income tax prior to beginning the cost segregation engagement. Collecting this information should be part of the pre-qualifying process. Some taxpayers will be subject to limitations on deductions that do not result in reduced taxes in the current year.

Net operating losses, passive activity loss limitations, and at-risk limitations are some of the possible limitations that may apply. These limitations are outside the scope of this text, and are the primary reason a tax professional is needed on the cost segregation "team." Engagement planning, execution, and reporting are all very important parts of the cost segregation process.

Calculating the Depreciation Deduction

There are many different ways to calculate depreciation for tax purposes. This chapter focuses on the method most commonly used by taxpayers, the Modified Accelerated Cost Recovery System (MACRS). MACRS is comprised of two subsystems: the General Depreciation System (GDS) and the Alternative Depreciation System (ADS). Of the two, GDS is most often used. Certain other methods are available, at the election of the taxpayer.

Since September 11, 2001, several tax incentives have been passed to encourage investment in new capital. These include a 30% bonus depreciation, a 50% bonus depreciation, and increased Section 179 expensing deductions. In September of 2003, the U.S. Treasury Department issued significant regulations relating to these deductions. These allowances affect the basis used for calculating depreciation and the total deduction for the year, which will be discussed later in this chapter.

Taxpayers with acquisitions that qualify for the bonus or additional first-year depreciation are typically required to take the deduction unless they formally elect not to when filing their taxes. *(See the discussion of Bonus Depreciation later in this chapter.)*

The General Depreciation System (GDS)

Depreciation deductions must be calculated according to the methods provided in IRC Section 168(a), which states:

> General rule. Except as otherwise provided in this section, the depreciation deduction provided by Section 167(a) for any tangible property shall be determined by using:
>
> 1. The applicable depreciation method,
> 2. The applicable recovery period, and
> 3. The applicable convention.

The applicable method is further described as:

1. The 200% declining balance method, switching to straight-line in the first year that it provides for a larger deduction. This method is used when the 150% declining balance or straight-line methods are not required or otherwise elected.[22]
2. The 150% declining balance method, switching to straight-line, required for 15- or 20-year property, certain farm property, and property for which a taxpayer makes an election under Section 168 (b)(2)(c) to have the 150% declining balance method apply.[23]
3. The straight-line method is used for:
    a. Non-residential real property
    b. Residential rental property
    c. Railroad grading or tunnel bore
    d. Property for which an election under Section 168 (b)(5) to use this method is made by the taxpayer[24]
    e. Trees or vines bearing fruit or nuts
    f. Water utility property or municipal sewers[25]

The applicable recovery period, as outlined in the IRC, is as follows:

- 3-year property: Class life of four years or fewer
- 5-year property: Class life of more than four years, but fewer than 10 years
- 7-year property: Class life of 10 or more years, but fewer than 16 years
- 10-year property: Class life of 16 or more years, but fewer than 20 years
- 15-year property: Class life of 20 or more years, but fewer than 25 years
- 20-year property: Class life of 25 or more years
- Water utility property: 25 years
- Residential rental property: 27.5 years
- Non-residential real property: 39 years[26]

The Applicable Convention: The final item needed for calculating the depreciation is the applicable convention. *Conventions* have the effect of treating the placed-in-service date for all assets as if it were the mid-point of the period in question. Sometimes called *averaging conventions,* they assume that most assets are purchased evenly throughout the year or set period. The general rule is that the half-year convention is used with respect to all property, unless it is non-residential real property, residential rental property, or a railroad grading or tunnel bore. A calendar-year taxpayer is treated as placing assets subject to the half-year convention in service at the mid-point of the year, or on June 30.

One exception to using the half-year convention is if the mid-quarter convention applies to assets placed in service during the year. A 40% test is used relative to assets placed in service in the fourth quarter of the year. Taxpayers compare the depreciable basis of assets depreciated at accelerated rates and purchased in the fourth quarter to all assets depreciated at accelerated rates. If the result is greater than 40%, the mid-quarter convention applies to all assets using accelerated methods.

A calendar-year taxpayer, in this case, is treated as placing the asset in service in the middle of the quarter. January through March purchases are considered as placed in service on February 15, and so on.[27] Assets that are acquired and disposed of (sold, transferred, stolen, ruined, and so forth) in the same year, and those that are subject to an election under IRC Section 179 (relating to expensing assets) are not included in the computation of the mid-quarter test.

Non-residential real property, residential rental property, and railroad property use the mid-month convention. The assumed placed-in-service date is the middle of the month or the 15th day of the month.[28]

How to Calculate Depreciation

The basis for depreciation is the adjusted basis of property used for determining gain or loss on the disposal. The original basis of an asset is generally the cost to acquire it.[29]

Calculation example:

> *Assume an asset has a basis for depreciation of $1,000, and a class life of 6 years. The asset is placed in service during the month of May 20X1. The asset is 5-year property because the class life is in the range of 4–10 years. Five-year property is depreciated under the following conditions:*
>
> - *The applicable method is the 200% declining balance method*
> - *The applicable recovery period is five years*
> - *The applicable convention is the half-year convention, assuming no other assets are placed in service during the year, and the 40% is not met*

To calculate the depreciation deduction, the following formula can be used for the initial year:

(basis/life in years x 200%)/2

Dividing by two results in a half-year allowance under the half-year convention.

The second and following years use a similar formula, except that the basis for depreciation is reduced first by the depreciation taken in prior years.[30] The half-year convention adjustment is not made again. Also, the result must be higher than the straight-line rate. The formula is stated as:

((basis-prior years depreciation)/life in years) x *200%*

Year 1

$1,000/5 years x *200% = $400*

$400/2 = $200

Year 2

($1,000 - $200)/5 years x *200%*

$800/5 years x *200%*

$160 x *200% = $320*

Year 3

($1,000 - $200 - $320)/5 years x *200%*

$480/5 years x *200%*

$96 x *200% = $192*

The calculations for years 4, 5, and 6 would be similar, except that the switch to straight-line depreciation method comes into play. Straight-line, for this purpose, is calculated by using the remaining life of the asset, taking into account the half-year in the final year of the depreciable life of the asset. Calculate the deduction both ways, and use the calculation that results in a higher deduction.

Year 4

($1,000 - $200 - $320 - $192)/5 years x *200%*

$288/5 years x *200%*

$57.60 x *200%= $115.20*

Or:

$288/remaining life

$288/2.5 years = $115.20

Since the results are equal, the deduction for Year 4 is $115.20

Year 5

($1,000 - $200 - $320 - $192 - $115.20)/5 years x *200%*

$172.80/5 years x *200%*

$34.56 x *200% = $69.12*

Or:

$172.80/1.5 years = $115.20

Since the straight-line calculation results in a higher deduction, that amount is used for the deduction for Year 5: $115.20.

Year 6

Year 6 will be the final year for a 5-year asset. The rate of depreciation to apply in this year is 100% of the remaining basis, or $57.60 in this case.

IRS Tables

The IRS guidance in Revenue Procedure 87-56 includes tables for use in calculating depreciation deductions under varying circumstances. The previous example calculating depreciation on 5-year property, using the 200% declining balance method and a half-year convention, could be calculated using Table A-1 in the Revenue Procedure.[31] Figure 2.1, a portion of Table A-1, is excerpted on the following page. (The complete table can be found at the end of the chapter.)

One would identify the type of property (e.g., 3-year or 5-year) and the year of the asset's life, and then use that percentage multiplied by the original basis of the asset to calculate the depreciation amount for that year. For the first year, the percentage would be 20%. $1,000 x 20% = $200.

Half-Year Convention			
Year	3-year property	5-year property	7-year property
1	33.33%	20%	14.29%
2	44.45%	32%	24.49%
3	14.81%	19.2%	17.49%
4	7.41%	11.52%	12.49%
5		11.52%	8.93%
6		5.76	8.92%
7			8.93%
8			4.46%

Figure 2.1

Additional tables are available for calculating depreciation when different conventions are needed.

Table A-2: Applied to assets using declining balance methods (200% and 150%) and the mid-quarter convention for assets placed in service in the *first quarter of the year.*

Table A-3: Applied to assets using declining balance methods (200% and 150%) and the mid-quarter convention for assets placed in service in the *second quarter of the year.*

Table A-4: Applied to assets using declining balance methods (200% and 150%) and the mid-quarter convention for assets placed in service in the *third quarter of the year.*

Table A-5: Applied to assets using declining balance methods (200% and 150%) and the mid-quarter convention for assets placed in service in the *fourth quarter of the year.*

Computer Software

Most practitioners use software to calculate depreciation. It is important to understand the software's terms (or vocabulary, like "personal property," "leasehold improvement," or "qualifying property") and how they will affect the output. Calculations should be double-checked independent of the software.

Basis Adjustments

When there are basis adjustments, they should be placed in the formula as if they were prior-year depreciation deductions. Once a basis adjustment is made, the IRS tables can no longer be used, and the formula calculation is necessary.

Basis adjustments may be required as a result of the operation of other provisions in tax law. One example is when the agreed purchase price of a building changes within a year following the acquisition (e.g., because of purchase and sales agreement, continuity of tenant leases, income targets, etc.), and where debt relating to the acquisition is reduced as a result of the decrease in price.

Cancellation of debt is normally considered income, however it may be excluded if the basis of the asset acquired is reduced by the amount of the "purchase price reduction."[32]

Non-Residential Real Property

Non-residential real property is defined generally as depreciable real property that is not residential rental property. According to the IRC:

> *The term non-residential real property means Section 1250 property, which is not residential rental property or property with a class life of less than 27.5 years.*

Non-residential real property is depreciated under the straight-line method over 39 years. The mid-month convention is used in the months of acquisition and disposal.

Section 1250 property is defined in Section 1250(c) as real property that is not Section 1245 property[33] and has been subject to an allowance for depreciation under Section 167. Section 1245 property does not include land and permanent improvements to land, such as buildings and their structural components. Land is not subject to the allowance for depreciation.[34] Land improvements are sometimes designated a class life range of 20–25 years and are also excluded from the definition of non-residential real property.

Non-residential real property includes buildings and their structural components used in a taxpayer's trade or business or for the production of income (rental operation) that are subject to exhaustion or wear and tear or obsolescence and that are not residential rental real property as described in Section 168(e)(2)(A).

Residential Rental Property

Residential rental property refers to any building or structure whose gross rental income is 80% or more from dwelling units. *Dwelling units* are houses or apartments used to provide living accommodations in a building or structure. Motels, hotels, or other property with more than one half of the units used on a transient basis are not considered residential rental real estate.[35] Gross rental income for purposes of this calculation includes the fair market value of the rental of owner-occupied space.

Residential rental property is depreciated over 27.5 years using the straight-line method and the mid-month convention. The following example illustrates the use of the mid-month convention.

A non-residential building that is real property is purchased for $500,000. The building is ready for use and placed in service on April 5, 2005. The building will be depreciated over 39 years, which is the recovery period assigned to non-residential real estate.[36] The applicable method is the straight-line method.[37] The applicable convention is the mid-month convention.[38]

($500,000/life in years, or 39) = $12,820.52, annual depreciation

(Annual depreciation)/number of months in the year = $1,068.38

(Months in service, including 1/2 month for the first month) x (the monthly depreciation amount) = the depreciation for the first year

8.5 months x $1068.38 = $9081.23

Years 2-39 deduction: the full year amount, or $12,820.52

In year 40, the remaining basis is taken.[39] 3.5 months x $1068.38 = $3739.01

Figure 2.2, table A-7a, is used for calculating depreciation on non-residential real property.

Month placed in service in the first year				
Year	January	February	March	April
1	2.461%	2.247%	2.033%	1.819%
2-39	2.564%	2.564%	2.564%	2.564%
40	.107%	.321%	.535%	.749%

Figure 2.2

There is a significant difference between the amount of depreciation deduction available under straight-line rates (with longer lives of 27.5 or 39 years) than if the same asset's deduction was calculated using shorter periods (5, 7, or 15 years) and accelerated rates. The following chart summarizes the depreciation of an asset by year for 5-, 7-, 15-, 27.5-, and 39-year property. Note the differences in earlier years' deductions—since the primary reason for performing cost segregation studies is enabling taxpayers to take a larger deduction earlier, in cases where Tangible Property or Land Improvement classifications can be supported.

Consider the asset with a depreciable basis of $100,000, and a January placed-in-service date. Depreciation deductions are shown in Figure 2.3—by year and property type for a $100,000 asset, and no bonus depreciation.

The practitioner should apply the taxpayer's expected tax rates to the deductions to show how the client's tax liabilities are affected by the reclassification of assets. When a change in accounting method is requested, the adjustment related to the change should be factored into any analysis. A further analysis of cash flows (without a "change in method" adjustment) can show the value of the change in tax liability in terms of today's dollars.

Figure 2.4 shows depreciation deductions by year and property type for a $100,000 asset, qualifying for the 30% Additional First-Year Depreciation Allowance.

Figure 2.5 shows depreciation deductions by year and property type for $100,000 asset qualifying for the 50% Additional First Year Depreciation Allowance.

Year	5-year property	7-year property	15-year property	27.5-year property	39-year property
1	$20,000	$14,290	$5,000	$3,485	$2,461
2	32,000	24,490	9,500	3,636	2,564
3	19,200	17,490	8,550	3,636	2,564
4	11,520	12,490	7,700	3,636	2,564
5	11,520	8,930	6,930	3,636	2,564
6	5,760	8,920	6,230	3,636	2,564
7		8,930	5,900	3,636	2,564
8		4,460	5,900	3,636	2,564
9			5,910	3,636	2,564
10			5,900	3,637	2,564
11			5,910	3,636	2,564
12			5,900	3,637	2,564
13			5,910	3,636	2,564
14			5,900	3,637	2,564
15			5,910	3,636	2,564
16			2,950	3,637	2,564
17				3,636	2,564
18				3,637	2,564
19				3,636	2,564
20				3,637	2,564
21				3,636	2,564
22				3,637	2,564
23				3,636	2,564
24				3,637	2,564
25				3,636	2,564
26				3,637	2,564
27				3,636	2,564
28				1,970	2,564
29					2,564
30					2,564
31					2,564
32					2,564
33					2,564
34					2,564
35					2,564
36					2,564
37					2,564
38					2,564
39					2,564
40					107

Figure 2.3

Year	5-year property	7-year property	15-year property	27.5-year property	39-year property
1	$44,000	$40,003	$33,500	$3,485	$2,461
2	22,400	17,143	6,650	3,636	2,564
3	13,440	12,243	5,985	3,636	2,564
4	8,064	8,743	5,390	3,636	2,564
5	8,064	6,251	4,851	3,636	2,564
6	4,032	6,244	4,361	3,636	2,564
7		6,251	4,130	3,636	2,564
8		3,122	4,130	3,636	2,564
9			4,137	3,636	2,564
10			4,130	3,637	2,564
11			4,137	3,636	2,564
12			4,130	3,637	2,564
13			4,137	3,636	2,564
14			4,130	3,637	2,564
15			4,137	3,636	2,564
16			2,065	3,637	2,564
17				3,636	2,564
18				3,637	2,564
19				3,636	2,564
20				3,637	2,564
21				3,636	2,564
22				3,637	2,564
23				3,636	2,564
24				3,637	2,564
25				3,636	2,564
26				3,637	2,564
27				3,636	2,564
28				1,970	2,564
29					2,564
30					2,564
31					2,564
32					2,564
33					2,564
34					2,564
35					2,564
36					2,564
37					2,564
38					2,564
39					2,564
40					107

Figure 2.4

Year	5-year property	7-year property	15-year property	27.5-year property	39-year property
1	$60,000	$57,145	$52,500	$3,485	$2,461
2	16,000	12,245	4,750	3,636	2,564
3	9,600	8,745	4,275	3,636	2,564
4	5,760	6,245	3,850	3,636	2,564
5	5,760	4,465	3,465	3,636	2,564
6	2,880	4,460	3,115	3,636	2,564
7		4,465	2,950	3,636	2,564
8		2,230	2,950	3,636	2,564
9			2,955	3,636	2,564
10			2,950	3,637	2,564
11			2,955	3,636	2,564
12			2,950	3,637	2,564
13			2,955	3,636	2,564
14			2,950	3,637	2,564
15			2,955	3,636	2,564
16			1,475	3,637	2,564
17				3,636	2,564
18				3,637	2,564
19				3,636	2,564
20				3,637	2,564
21				3,636	2,564
22				3,637	2,564
23				3,636	2,564
24				3,637	2,564
25				3,636	2,564
26				3,637	2,564
27				3,636	2,564
28				1,970	2,564
29					2,564
30					2,564
31					2,564
32					2,564
33					2,564
34					2,564
35					2,564
36					2,564
37					2,564
38					2,564
39					2,564
40					107

Figure 2.5

The Alternative Depreciation System (ADS)

ADS is one of two systems used to calculate depreciation under MACRS. It is required for certain types of taxpayers, and, in some cases, a taxpayer may elect to use it. The terminology and classification issues are similar to the GDS system. The Internal Revenue Code, which addresses these differences, outlines which taxpayers are required to use this system of calculating depreciation:

> *Section 168(g) provides in the case of certain taxpayers the depreciation deduction provided by Section 167(a) shall be determined under the Alternative Depreciation System.*

This requirement applies to the following situations:

- Any tangible property that, during the taxable year, is used predominantly outside the U.S.
- Any tax-exempt use property
- Any tax-exempt bond-financed property
- Any imported property covered by an executive order under Paragraph 6
- Any property to which an election under Paragraph 7 applies

The deduction is computed by using:

1. The straight-line method
2. The applicable convention (same as GDS System)
3. The recovery period determined for ADS in Section 168(g)(2)(C)

The recovery periods for ADS are included in Publication 946, in the same table as the General Depreciation System. (The complete tables can be found in the discussion of asset classification in Chapter 5.)

Taxpayers may elect this method for any one or more classes of property. The election relating to residential rental property and non-residential real property may be made on a property-by-property basis. The election is made annually, and once made is irrevocable.

Practitioners should be aware of ADS and factor it into any planning considerations for clients who use it. The benefits of cost segregation, although less dramatic, are still there. It is also important to note that in some exempt entities, the amount of current funding available may be determined by the amount of depreciation allowance in a given year or period of years. This effect on cash flow planning may make cost segregation attractive in these cases. The benefits must be quantified differently under these circumstances.

Bonus Depreciation

The Job Creation and Worker Assistance Act of 2002 created a 30% additional first-year depreciation allowance for qualifying MACRS property under Section 168(k), also referred to as the "Bonus Depreciation." Section 168(k) was modified in 2003 to increase the bonus to 50%. Property that qualifies for the bonus will be based on the acquisition date requirements in the code. The Treasury Department issued extensive regulations dealing with the Bonus Depreciation allowance under both the 30% and 50% rules. Generally, the only distinguishing requirement between the two is the date of acquisition.

The 30% Additional First Year Depreciation is allowed for property that meets the following requirements:

1. The initial use of the property begins with the taxpayer.
2. The property is acquired after 9/10/2001 and before 9/11/2004. Property acquired after 9/10/2001 under a binding written contract entered into prior to 9/11/2001 does not qualify.
3. To qualify, the property must be:
    a. Depreciable[40] under MACRS and have a recovery period of 20 or fewer years. Assets required to be depreciated under ADS do not qualify.[41]
    b. MACRS water utility property,
    c. Computer software depreciable over three years under Section 167, or
    d. "Qualified leasehold improvement property."

The bonus depreciation is allowed in full for the Alternative Minimum Tax (AMT). There is no AMT adjustment required for assets that qualify for bonus, and for which the bonus is actually taken (no election out of Section 168(k)). The taxpayer may elect for any one or all classes of qualifying property not to have the provisions of Section 168(k) apply. It is important to note that the allowance under Section 168(k) is mandatory unless the "election out" is made. Since the bonus depreciation is treated as a deduction for depreciation, it is subject to recapture as ordinary income under Section 1245 when the property is sold at a gain.

Qualified Leasehold Improvement Property

This category of depreciable property includes improvements to the interior of a building that is non-residential real property.

The following conditions must apply:

1. Improvements made by a lessor, lessee, or sub-lessee pursuant to a lease.
2. Improvements placed in service more than three years after the building was first placed in service (e.g., the building must be more than three years old).
3. The lessor and the lessee may not be related.

Expenditures for the following do not qualify:

- Enlargement of a building
- Elevators or escalators
- Any structural component that benefits a common area or the internal structural framework of a building

50% Additional First-Year Depreciation

The 50% allowance is determined under the same provisions in effect for the 30% deduction, except the acquisition must take place after 5/6/2003, and no binding written contract must have been entered into before that date. The asset must be placed in service by 12/31/2004.

Calculation example:

> *Office furniture, considered 7-year property, is purchased at a cost of $1,025. It is placed in service at the time of purchase, February 12, 2003, and no contract was in place before September 11, 2001. The calculated 30% bonus depreciation amount is $308 ($1,025 x 30%). The amount remaining after the bonus is used to calculate the regular depreciation deduction using the tables for 7-year property with the half-year convention is $1,025 – $308 = $717. The rate in the table for the first year is 14.29%. Multiplying this rate by the depreciable basis for year one results in a $102 regular depreciation. Combine the two calculated amounts to obtain the total depreciation deductions for the first year ($308 bonus depreciation + $102 regular depreciation = $410).*

The newly issued regulations provide examples of a binding written contract and define it as one that is enforceable under state law against the taxpayer (or a predecessor). Damages are not limited to a specified amount.

The regulations also provide that a contract is binding even if it is subject to a condition, as long as the condition is not within the control of either one of the parties or a predecessor. Further, an option to either acquire or sell property is not treated as a binding contract. A binding contract does not include a supply or similar agreement if the amount and design of the property to be purchased have not been specified. In this case, the contract is not treated as binding until both the amount and design specifications are specified.[42]

Self-Constructed Property

There are special rules for this type of property relating to the bonus depreciation deduction and binding written contracts. Self-constructed property may include buildings built under contract. The regulation states that property that is manufactured, constructed, or produced for a taxpayer by another person under a binding written contract entered into prior to construction for use by the taxpayer in trade, business, or for the production of income is considered to be manufactured, constructed, or produced by the taxpayer. The taxpayer must meet the acquisition date requirements if construction begins after the applicable dates.

When Does Construction Begin?

Construction is considered to commence when physical work of a significant nature gets under way. Physical work does not include preliminary activities, such as planning or designing, securing financing, establishing a location, or researching. Construction cannot be considered as commencing until more than 10% of the cost is incurred (or paid in the case of a cash-basis taxpayer). These costs do not include the cost of land or the above-mentioned preliminary activities.

"Accrual Methods" versus "Cash Methods"

A cash method taxpayer incurs a cost when paying cash. An accrual method taxpayer incurs cost when it is possible to measure the amount of the expenditure and determine that an amount is required to be paid at some point in the future.

Components of a larger constructed asset are analyzed separately to determine if there was a binding written contract in place before the applicable dates related to bonus depreciation. In the case of an acquired component, the acquisition date requirements need to be met in order for the component asset to qualify for a 30% bonus depreciation deduction, or for the 50% bonus depreciation deduction. In the case of a constructed component, the same rules apply as those for the larger constructed asset. If a component does not qualify due to the acquisition date rules, it will not disqualify the larger constructed asset from the bonus deductions. The basis of the larger constructed asset does not include the basis of the non-qualifying component.

Alternatively, if a larger component does not meet the acquisition date requirements for the bonus depreciation allowance, then no component will qualify for the bonus allowance. The assets must be in place by January 1, 2005, under the rules for bonus depreciation deduction. When a larger constructed asset is not placed in service by January 1, 2005, any component meeting the acquisition requirements under a binding written contract entered into before January 1, 2005 will qualify, assuming all other requirements are met.

Examples: The following examples illustrate the acquisition date requirements that need to be met in order for a component asset to qualify for the 30% or the 50% bonus depreciation deduction, according to Section 1.168(k)-1T (4)(v).[43] (Note that while the example taxpayers are commercial businesses, these rules also apply to individuals and other entities, such as trustees of a trust.)

Example 1: *On September 1, 2001, Corporation J entered into a written agreement with Manufacturer K, to purchase 20 new lamps for $100 each within the next two years. Although the agreement specifies the number of lamps to be purchased, it does not stipulate their design. Accordingly, the agreement is not a binding contract pursuant to paragraph (b)(4)(ii)(D) of Section 1.168(k)-1T.*

Example 2: *Using the same basic facts in Example 1, on December 1, 2001, Corporation J placed a purchase order with Manufacturer K to purchase 20 new model XPC5 lamps for $100 each for a total amount of $2,000. Because the agreement specifies the number of lamps to be purchased, and the purchase order specifies the design of the lamps to be purchased, the purchase order placed by Corporation J with Manufacturer K on December 1, 2001, is a binding contract pursuant to paragraph (b)(4)(ii)(D). Accordingly, the cost of the 20 lamps qualifies for the 30% additional first-year depreciation deduction.*

Example 3: *Using the same basic facts in Example 1, assume the written agreement between Corporation J and Manufacturer K is, instead, to purchase 100 model XPC5 lamps for $100 each within the next two years. Because this agreement specifies the amount and design of the lamps to be purchased, the agreement is a binding contract pursuant to paragraph (b)(4)(ii)(D). Accordingly, because the agreement was entered into before September 11, 2001, any lamp acquired by Corporation J under this contract does not qualify for the additional first year depreciation.*

Example 4: *On September 1, 2001, assume Company L began constructing an electric generation power plant for its own use. On November 1, 2002, the company ceased construction of the power plant prior to its completion. Between September 1, 2001 and November 1, 2002, Company L incurred $3,000,000 for the* *construction of the power plant. On May 6, 2003, the company resumed construction of the power plant and completed its construction on August 31, 2003. Between May 6, 2003 and August 31, 2003, Company L incurred another $1,600,000 to complete the construction of the power plant and, on September 1, 2003 placed the plant in service. None of Company L's total expenditures of $4,600,000 qualify for the additional first-year depreciation deduction, because pursuant to (b)(4)(iii)(A), the company began constructing the plant before September 11, 2001.*

Example 5: *Using the same basic facts in Example 4, assume that Company L began constructing the plant (for its own use) on October 1, 2001. The company's total expenditures of $4,600,000 would qualify for the additional first-year depreciation deduction, because it began constructing the power plant after September 10, 2001 and placed the plant in service before January 1, 2005. Accordingly, the additional first-year depreciation deduction for the power plant would be $1,380,000 (30% of $4,600,000).*

Example 6: *On August 1, 2001, Company M entered into a written binding contract to acquire a new turbine as a component part of a new electric generation power plant that is being constructed on Company M's behalf. The construction of the new electric generation power plant commenced in November, 2001, and was completed in November, 2002. Because Company M entered into a written binding contract to acquire a component part (the new turbine) prior to September 11, 2001 pursuant to paragraph (b)(4)(iii)(C), the component part does not qualify for the additional first year depreciation deduction. However, pursuant to paragraphs (b)(4)(iii)(A) and (C), the new plant constructed for Company M will qualify for the 30% additional first-year depreciation deduction, because construction on the new plant began after September 10, 2001 and before May 6, 2003. Accordingly, the unadjusted depreciable basis of the new plant that is eligible for the 30% additional first-year depreciation deduction must not include the unadjusted depreciable basis of the new turbine.*

Example 7: *Using the same basic facts in Example 6, assume that Company M entered into a binding written contract to acquire the new turbine on September 30, 2002, and construction of the new plant commenced on August 1, 2001. Because Company M began construction prior to September 11, 2001, pursuant to paragraphs (b)(4)(iii)(A) and (C), neither the new plant constructed for Company M nor the turbine will qualify for the additional first-year depreciation deduction because self-construction began prior to September 11, 2001.*

Example 8: *On September 1, 2001, Company N began constructing property for its own use. On October 1, 2001, Company N sold its rights in the property to Company O, a related party under IRC Section 267(b). Pursuant to paragraph (b)(4)(iv) of this section, the property is not eligible for the additional first-year depreciation because Companies N and O are related parties and the construction of the property began before September 11, 2001.*

Example 9: *On September 1, 2001, Company P entered into a written binding contract to acquire property. On October 1, 2001, Company P sold its rights to the* *property to Company Q, a related party under IRC Section 267(b). The property is not eligible for the additional first-year depreciation deduction because Companies P and Q are related parties and a binding written contract for the acquisition of property was in effect prior to September 11, 2001.*

Example 10: *Prior to September 11, 2001, company R began constructing an electric generation power plant for its own use. On May 1, 2003, prior to the completion of the power plant, Company R transferred the rights to own and use this power plant to Company S, an unrelated party, for $6,000,000. Between May 6, 2003 and June 30, 2003, S, a calendar-year taxpayer, incurred an expense of $1,200,000 to complete the construction of the power plant. On August 1, 2003, Company S placed the power plant in service. Because Companies R and S are not related parties, their transaction will not be disqualified under Section 1.168(k)-1T. Accordingly, Company S's total expenditures of $7,200,000 for the power plant qualify for the additional first-year depreciation deduction. Company S's additional first-year depreciation deduction for the power plant will be $2,400,000, (30% of $6,000,000, plus 50% of $1,200,000).*

The $6,000,000 portion of the $7,200,000 unadjusted depreciable basis qualifies for the 30% additional first-year depreciation deduction, because that portion of the total unadjusted depreciable basis was acquired by S after September 10, 2001, and before May 6, 2003. However, because Company S began construction to complete the power plant after May 5, 2003, the $1,200,000 portion of the total $7,200,000 unadjusted depreciable basis qualifies for the 50% additional first-year depreciation deduction.

Tables

IRS Publication 946 tables may be used to calculate depreciation instead of formula calculations. Find the column that has the life of the asset and the row corresponding to the year for which you would like to calculate depreciation. The decimal is multiplied by the original depreciable basis of the asset.

Be sure to use the correct table. Table A-1 will be used most of the time. Tables A-2 through A-5 are used if you have met the mid-quarter test. Tables A-6, A-7, and A-7a are used for buildings in general (buildings 27.5 or 39 years). Other tables are available if you are using a different method (150% declining balance or straight line) or a non-standard MACRS life. *(See the end of the chapter for the IRS tables).*

Change in Accounting Method

In this chapter, we have discussed the many facets of depreciating property that is used in a trade or business for the production of income. A cost segregation study can identify opportunities for immediate tax deductions or accelerated tax depreciation deductions that would not otherwise be available until later years. When cost segregation studies are performed on facilities that were constructed or acquired in the past (as far back as 1987), the taxpayer must file *a request for consent to change accounting methods* with the IRS in order to take advantage of the benefits.

Accounting Method Adoption

The IRS assumes adoption of a particular method of accounting if you treat an item in the same manner on two consecutively filed tax returns. If, on the initial return, you calculate depreciation using a certain method, that method is "adopted" when you file your return in year two, using the same method. (Where an item is treated incorrectly on a return, two consecutive returns must be filed using that method in order to file for a change of accounting method. However, it is not necessary for two returns to be filed before the change of accounting method is adopted if the original, first year method is correct.) In most cases, to change the method of accounting, you need the consent of the commissioner,[44] which is obtained by filing Form 3115.

Nature of the Change in Method

The change in accounting method that is required in order to reclassify depreciation deductions is a change from an "impermissible" method to a "permissible" method. A taxpayer who had not claimed the correct amount of depreciation deductions will now claim the correct amount. The change in method is reported to the IRS on Form 3115 during the year of change.[45]

When a taxpayer files the form in accordance with the provisions of Revenue Procedure 99-49 or Revenue Procedure 2002-9, the consent of the commissioner is automatic. A change in accounting method may not be made if the same change has been made within the five preceding years, including the year of change, whether or not the change was granted.

Section 481(a) Adjustment

Section 481(a) Adjustment is the amount of change in depreciation that is allowable from the amount of the depreciation taken prior to the year of the change of accounting method. When the cost segregation study is complete, the assets identified are depreciated using the methods and periods as corrected. The result is compared to the depreciation deductions claimed on all prior returns for the same assets. Revenue Procedure 99-49 required that the adjustment, whether positive or negative, be included in the income or deducted in arriving at taxable income equally over a period of four years, although revenue ruling 2002-19 changed that four-year rule to one year for adjustments that are in the taxpayer's favor. This benefits the taxpayer, as it creates a negative adjustment to income, or a deduction. The taxpayer may take into account the entire amount of income in the year of change by making an entry on the appropriate line of Form 3115.[46]

Timely Filing

A change in accounting method application should generally be filed within the first 180 days of the year of the change. This requirement is waived for applications filed in accordance with Revenue Procedure 99-49—which must be filed in duplicate. The original must be filed with the return for the year of change by the due date (including extensions), and a copy must be sent to the national office of the IRS no earlier than the first day of the year of change, and no later than the date the original is filed with the return for the year of change. There is no user fee for an application filed under Revenue Procedure 99-49.

Special Statement for Change of Property Class

The Revenue Procedure requires a specific statement be attached to the application when changing the classification of assets from Section 1250 to Section 1245. The statement must include a description of the facts, the relevant law supporting the change, and a representational statement. The statement may read as follows:

> *"Each item of property that is the subject of the application filed under Section 2.01 of the Appendix of Revenue Procedure 99-49 for the year of change beginning 1/1/2001, and that is reclassified from non-residential real property to an asset class of Revenue Procedure 87-56, 1987-2 C.B. 674, that does not explicitly include Section 1250 property, is Section 1245 property for the purposes of depreciation."*[47]

Further Developments

Several cases involving changes in accounting method have been litigated successfully with the IRS. The point of the litigation was to determine whether the taxpayer had a change in accounting method (the IRS position) for a change in useful life without a change in the underlying facts and circumstances. Circuit courts had found differently for different taxpayers, which led the IRS to remediate the situation.

The IRS issued new regulations and revenue procedures in late 2003 and early 2004. The new regulations stated that a change in useful life without a change in the underlying facts and circumstances is a change in accounting method—which requires consent of the commissioner. For the years ending prior to December 30th, 2003, the IRS gives the taxpayer the opportunity to file either amended returns or a Form 3115 to request a change in accounting method. For years after December 30th, 2003, Form 3115 must be filed. This applies to the types of changes involved in cost segregation.

As mentioned earlier, one of the conditions or rules that determines whether a taxpayer has adopted a method of accounting is the use of the same method on two consecutively filed tax returns. However, a new Revenue Procedure (2004-11) has been issued by the IRS in the interest of lightening the burden on taxpayers who wish to change their method of accounting (where two returns have not been filed). The IRS will allow taxpayers to use Form 3115 instead of filing amended tax returns to claim the correct amount of depreciation in the year a disposition occurs.

For more information, refer to the following cases and rulings:

- Brookshire Brothers Holding 2003-1 USTC ¶50,214
- Kurzet 2000-2 USTC ¶50,671
- Green Forest Manufacturing TC MEMO 2003-75
- Chief Counsel Notice CCN 2004-007 and 2004-024

Sample Estimates

Figures 2.6–2.9 are sample estimates of the tax benefits for a retail strip plaza, manufacturing facility, hotel, and apartment complex. Each worksheet shows estimated depreciation and the net present value of the tax benefits before and after a cost segregation study. These estimates are typically communicated to clients in the feasibility stage. This format (with final depreciation established from the study) may also be included in the final cost segregation reports, or may be used as a tool to communicate the results to clients.

The apartment complex is residential real property (27.5-year property). No bonus depreciation was taken. The manufacturing facility is non-residential (39-year property), and a 50% bonus depreciation was taken on qualifying property. The hotel is 39-year property, and a 30% bonus depreciation was taken on qualifying property.

The hotel schedule also includes a "Section 481 adjustment" due to a change in accounting method. The strip plaza placed in service in 2000 includes a "Section 481 adjustment." Bonus depreciation was not available since the assets were placed in service prior to September 11, 2001. Each estimate has the same discount rates, federal tax rates, and state tax rates, although these differ across the country.

Conclusion

Depreciation deduction is the central concept in cost segregation analysis. Cost segregation studies are performed to maximize the amount of assets that may be classified as personal property or land improvements, and that may therefore be eligible for accelerated depreciation under MACRS. Practitioners must have a complete understanding of depreciation calculations, rules, and methods in order to assess—and articulate to clients—the benefits of a cost segregation study.

The next requirement is understanding the law and analysis behind the classification of property for depreciation. Chapter 3 reviews the legal framework, Chapter 4 provides case summaries and significant rulings, and Chapter 5 addresses classification of property.

[1] Internal Revenue Code 1(a)-(d) imposes tax on individuals based on marital status and qualifying dependents. 1(e) imposes tax on estates and trusts.

[2] Internal Revenue Code 11(a)

[3] Income brackets refer to a range of income that is taxed at certain rates. For the 2002 tax year, a married individual filing a joint return with taxable income of between $46,700 and $112,850 was taxed at 27% of the income in that range. They are in the "27% bracket."

[4] Internal Revenue Code §63(a)

[5] Internal Revenue Code §61(a)

[6] By way of example, cancellation of indebtedness can be excluded if the taxpayer was insolvent at the time the debt was forgiven, was in a bankruptcy proceeding under Title 11, if the debt was qualified farm indebtedness or qualified real property business indebtedness. See §108(a)(1)(A)-(D)

[7] Internal Revenue Code §162 allows a deduction for "ordinary and necessary business expenses in carrying on a trade or business" and includes items like compensation, travel, and rent.

[8] Internal Revenue Code §163

[9] Internal Revenue Code §164

[10] Internal Revenue Code §165

[11] Internal Revenue Code §166

[12] Internal Revenue Code §167 and §168

[13] Internal Revenue Code §178 relating to the acquisition of a lease

[14] Internal Revenue Code §190

[15] Internal Revenue Code §198

[16] Internal Revenue Code §§169-197

[17] Internal Revenue Code, Subchapter A, Part IX. §§261-280H

[18] Internal Revenue Code §263A(b)(1)

[19] Internal Revenue Code §263A(a)(1)(B)

[20] Internal Revenue Code §263(a)(1)

[21] The deductions increase in the current year (or earlier years) results in decreased deductions in later years of the assets life. The later years can be over a period as long as 40 years away. It is important to remember the amount of the deduction is not changed, only the timing.

[22] IRC Section 168(b)(1)

[23] IRC Section 168(b)(2)

[24] The election under Section 168(b)(5) may be made for any one or more classes of property. The election applies to all property in the class for the tax year and the election is irrevocable.

[25] IRC Section 168(b)(3)

[26] IRC Section 168(c) & (e) provide for the classification of property and the recovery period.

[27] IRC Section 168(d)(3)

[28] IRC Section 168(d)(2)

[29] IRC Sections 167(g), 1011, and 1012

[30] It should be noted that this is a very simple example and does not consider other adjustments to basis that may be required by other sections of the IRC, such as basis reductions under purchase price reductions, Section 179 deductions, or other cancellation of debt scenarios.

[31] Tables for depreciation are included in IRS Publication 946, which reproduces the Revenue Procedure 87-56. The tables cover assets form 3-20 year lives.

[32] IRC Section 108

[33] The definition of Section 1245 property is found in 1245(a)(3).

[34] Treasury Regulations 1.167(a)(2).

[35] Transient basis is described as periods of use which typically do not exceed 30 days.

[36] IRC Section 168(c).

[37] IRC Section 168(b)(3)(A).

[38] IRC Section 168(d)(2)(A).

[39] The final year is often adjusted to account for rounding difference, since depreciation cannot exceed the depreciable basis of the underlying property.

[40] Section 197 intangibles do not qualify.

[41] An election by the taxpayer to use the ADS system does not disqualify the asset from the bonus depreciation allowance.

[42] Treasury Decision 9091, September 5, 2003.

[43] Treasury Decision 9091, September 5, 2003.

[44] See automatic change procedures in Revenue Ruling 99-49.

[45] Treasury Regulations Section 1.446-1(e)(3)(i).

[46] "The De-Minimis Rule" is elective.

[47] Revenue Procedure 99-49 Appendix Section 2.01(3)(b)(viii).

Sample Apartment Complex
Results of Cost Segregation Study
2004 Residential

Cumulative present value on taxes deferred at 8%
$345,870
Total taxes deferred 2004 - 2007
$499,932
Total additional depreciation claimed 2004-2007
$1,249,829

Results of Cost Segregation Study	Segregated
Building - 27.5 Yr S/L	$5,070,000
Land Improve - 15 Yr 150% DB	1,248,000
Furnishings - 7 Yr 200% DB	0
Equipment - 5 Yr 200% DB	1,482,000
Total Recovery Property	$7,800,000

Assumptions	
Discount Rate	8.0%
Effective Fed'L Tax Rate	33.0%
Effective State Tax Rate	7.00%

Year	(A) Total Original Cost Deprec	Depreciation Based on Segregated Cost Recovery: 27.5 Yrs	15 Yrs	7 Yrs	5 Yrs	(B) Total Segregated Cost Deprec	(B)-(A) **Additional Depreciation Computed**	**Total Taxes Deferred**	Sub-total By Year	Present Value of Cumulative Taxes Deferred
2004	177,294	115,241	62,400	0	296,400	474,041	**296,747**	118,699	**118,699**	109,906
2005	283,608	184,345	118,560	0	474,240	777,145	**493,537**	197,415	**197,415**	279,158
2006	283,608	184,345	106,704	0	284,544	575,593	**291,985**	116,794	**116,794**	371,873
2007	283,608	184,345	96,096	0	170,726	451,168	**167,560**	67,024	**$67,024**	421,137
2008	283,608	184,345	86,486	0	170,726	441,558	157,950	63,180	$63,180	464,136
2009	283,608	184,345	77,750	0	85,363	347,459	63,851	25,540	$25,540	480,231
2010	283,608	184,345	73,632	0	0	257,977	(25,631)	(10,252)	($10,252)	474,249
2011	283,608	184,345	73,632	0	0	257,977	(25,631)	(10,252)	($10,252)	468,710
2012	283,608	184,345	73,757	0	0	258,102	(25,506)	(10,202)	($10,202)	463,606
2013	283,686	184,396	73,632	0	0	258,028	(25,658)	(10,263)	($10,263)	458,852
2014	283,608	184,345	73,757	0	0	258,102	(25,506)	(10,202)	($10,202)	454,477
2015	283,686	184,396	73,632	0	0	258,028	(25,658)	(10,263)	($10,263)	450,401
2016	283,608	184,345	73,757	0	0	258,102	(25,506)	(10,202)	($10,202)	446,650
2017	283,686	184,396	73,632	0	0	258,028	(25,658)	(10,263)	($10,263)	443,155
2018	283,608	184,345	73,757	0	0	258,102	(25,506)	(10,202)	($10,202)	439,939
2019	283,686	184,396	36,816	0	0	221,212	(62,474)	(24,990)	($24,990)	432,645
2020	283,608	184,345	0	0	0	184,345	(99,263)	(39,705)	($39,705)	421,914
2021	283,686	184,396	0	0	0	184,396	(99,290)	(39,716)	($39,716)	411,975
2022	283,608	184,345	0	0	0	184,345	(99,263)	(39,705)	($39,705)	402,775
2023	283,686	184,396	0	0	0	184,396	(99,290)	(39,716)	($39,716)	394,254
2024	283,608	184,345	0	0	0	184,345	(99,263)	(39,705)	($39,705)	386,366
2025	283,686	184,396	0	0	0	184,396	(99,290)	(39,716)	($39,716)	379,061
2026	283,608	184,345	0	0	0	184,345	(99,263)	(39,705)	($39,705)	372,298
2027	283,686	184,396	0	0	0	184,396	(99,290)	(39,716)	($39,716)	366,035
2028	283,608	184,345	0	0	0	184,345	(99,263)	(39,705)	($39,705)	360,238
2029	283,686	184,396	0	0	0	184,396	(99,290)	(39,716)	($39,716)	354,868
2030	283,608	184,345	0	0	0	184,345	(99,263)	(39,705)	($39,705)	349,897
2031	248,196	161,327	0	0	0	161,327	(86,869)	(34,747)	($34,747)	**345,870**
Total	**$7,800,000**	**$5,070,000**	**$1,248,000**	**$0**	**$1,482,000**	**$7,800,000**	**($0)**	**($0)**	**($0)**	
% of Total	100.00%	65.00%	16.00%	0.00%	19.00%	100.00%				

Figure 2.6

Credit: IRS Publication 946, "How to Depreciate Property."

Sample Manufacturing Facility
Results of Cost Segregation Study
June 2004 Addition-HalfYr With 50% BONUS

Cumulative present value on taxes deferred at 8%
$176,898
Total taxes deferred 2004-2007
$231,781
Total additional depreciation claimed 2004-2007
$579,451

Results of Cost Segregation Study	Segregated
Building - 39 Yr	$1,630,000
Land Improve - 15 Yr 150% DB	420,000
Furnishings - 7 Yr 200% DB	390,000
Equipment - 5 Yr 200% DB	60,000
Total Recovery Property	$2,500,000

Assumptions	
Discount Rate	8%
Effective Fed'L Tax Rate	33.0%
Effective State Tax Rate	7.00%

	(A)					(B)	(B)-(A)			
		Depreciation Based on Segregated Cost Recovery								
Year	Total Original Cost Deprec	39 Yrs	15 Yrs	7 Yrs	5 Yrs	Total Segregated Cost Deprec	**Additional Depreciation Computed**	**Total Taxes Deferred**	Sub-total By Year	**Present Value of Cumulative Taxes Deferred**
2004	34,775	22,673	220,500	222,866	36,000	502,039	**467,264**	**186,905**	186,905	173,060
2005	64,100	41,793	19,950	47,756	9,600	119,099	**54,999**	**22,000**	22,000	191,922
2006	64,100	41,793	17,955	34,106	5,760	99,614	**35,514**	**14,206**	14,206	203,199
2007	64,100	41,793	16,170	24,356	3,456	85,775	**21,675**	**8,670**	8,670	209,572
2008	64,100	41,793	14,553	17,414	3,456	77,216	13,116	5,246	5,246	213,142
2009	64,100	41,793	13,083	17,394	1,728	73,998	9,898	3,959	3,959	215,637
2010	64,100	41,793	12,390	17,414	0	71,597	7,497	2,999	2,999	217,387
2011	64,100	41,793	12,390	8,697	0	62,880	(1,220)	(488)	(488)	217,123
2012	64,100	41,793	12,411	0	0	54,204	(9,896)	(3,959)	(3,959)	215,142
2013	64,100	41,793	12,390	0	0	54,183	(9,917)	(3,967)	(3,967)	213,305
2014	64,100	41,793	12,411	0	0	54,204	(9,896)	(3,959)	(3,959)	211,607
2015	64,100	41,793	12,390	0	0	54,183	(9,917)	(3,967)	(3,967)	210,032
2016	64,100	41,793	12,411	0	0	54,204	(9,896)	(3,959)	(3,959)	208,576
2017	64,100	41,793	12,390	0	0	54,183	(9,917)	(3,967)	(3,967)	207,225
2018	64,100	41,793	12,411	0	0	54,204	(9,896)	(3,959)	(3,959)	205,977
2019	64,100	41,793	6,195	0	0	47,988	(16,112)	(6,445)	(6,445)	204,096
2020	64,100	41,793	0	0	0	41,793	(22,307)	(8,922)	(8,922)	201,685
2021	64,100	41,793	0	0	0	41,793	(22,307)	(8,922)	(8,922)	199,452
2022	64,100	41,793	0	0	0	41,793	(22,307)	(8,922)	(8,922)	197,385
2023	64,100	41,793	0	0	0	41,793	(22,307)	(8,922)	(8,922)	195,470
2024	64,100	41,793	0	0	0	41,793	(22,307)	(8,922)	(8,922)	193,698
2025	64,100	41,793	0	0	0	41,793	(22,307)	(8,922)	(8,922)	192,057
2026	64,100	41,793	0	0	0	41,793	(22,307)	(8,922)	(8,922)	190,537
2027	64,100	41,793	0	0	0	41,793	(22,307)	(8,922)	(8,922)	189,130
2028	64,100	41,793	0	0	0	41,793	(22,307)	(8,922)	(8,922)	187,828
2029	64,100	41,793	0	0	0	41,793	(22,307)	(8,922)	(8,922)	186,621
2030	64,100	41,793	0	0	0	41,793	(22,307)	(8,922)	(8,922)	185,504
2031	64,100	41,793	0	0	0	41,793	(22,307)	(8,922)	(8,922)	184,470
2032	64,100	41,793	0	0	0	41,793	(22,307)	(8,922)	(8,922)	183,513
2033	64,100	41,793	0	0	0	41,793	(22,307)	(8,922)	(8,922)	182,626
2034	64,100	41,793	0	0	0	41,793	(22,307)	(8,922)	(8,922)	181,805
2035	64,100	41,793	0	0	0	41,793	(22,307)	(8,922)	(8,922)	181,045
2036	64,100	41,793	0	0	0	41,793	(22,307)	(8,922)	(8,922)	180,341
2037	64,100	41,793	0	0	0	41,793	(22,307)	(8,922)	(8,922)	179,689
2038	64,100	41,793	0	0	0	41,793	(22,307)	(8,922)	(8,922)	179,086
2039	64,100	41,793	0	0	0	41,793	(22,307)	(8,922)	(8,922)	178,527
2040	64,100	41,793	0	0	0	41,793	(22,307)	(8,922)	(8,922)	178,010
2041	64,100	41,793	0	0	0	41,793	(22,307)	(8,922)	(8,922)	177,531
2042	64,100	41,793	0	0	0	41,793	(22,307)	(8,922)	(8,922)	177,087
2043	29,425	19,185	0	0	0	19,185	(10,240)	(4,109)	(4,109)	**176,898**
Total	**$2,500,000**	**$1,630,000**	**$420,000**	**$390,000**	**$60,000**	**$2,500,000**	**($0)**	**$0**	**$0**	
% of Total	100.00%	65.20%	16.80%	15.60%	2.40%	100.00%				

Figure 2.7
Credit: IRS Publication 946, "How to Depreciate Property."

Sample Hotel
Results of Cost Segregation Study
March - Addition-Half Year With 30% BONUS

Cumulative present value on taxes deferred at 8%
$327,819

Total taxes deferred 2004-2007
$446,280

Total additional depreciation claimed 2004-2007
$1,115,705

Results of Cost Segregation Study	**Segregated**
Building - 39 Yr	**$2,650,000**
Land Improve - 15 Yr 150% DB	**775,000**
Furnishings - 7 Yr 200% DB	**0**
Equipment - 5 Yr 200% DB	**925,000**
Total Recovery Property	**$4,350,000**

Assumptions	
Discount Rate	8%
Effective Fed'L Tax Rate	33.0%
Effective State Tax Rate	7.00%

	(A)					(B)	(B)-(A)			
Year	Total Original Cost Deprec	Depreciation Based on Segregated Cost Recovery 39 Yrs	15 Yrs	7 Yrs	5 Yrs	Total Segregated Cost Deprec	**Additional Depreciation Computed**	**Total Taxes Deferred**	Sub-total By Year	**Present Value of Cumulative Taxes Deferred**
2003	88,436	53,875	259,625	0	407,000	720,500	632,064	0		0
Subtotal	**$88,436**	**$53,875**	**$259,625**	**$0**	**$407,000**	**$720,500**	**$632,064**	**$0**		**0**
2004 Catch-Up							**632,064**	**252,825**		
2004	111,534	67,946	51,538	0	207,200	326,684	**215,150**	**86,059**	338,884	313,781
2005	111,534	67,946	46,384	0	124,320	238,650	**127,116**	**50,846**	50,846	357,374
2006	111,534	67,946	41,773	0	74,592	184,311	**72,777**	**29,110**	29,110	380,482
2007	111,534	67,946	37,595	0	74,592	180,133	**68,599**	**27,440**	27,440	400,651
2008	111,534	67,946	33,798	0	37,296	139,040	27,506	11,002	11,002	408,139
2009	111,534	67,946	32,008	0	0	99,954	(11,581)	(4,633)	(4,633)	405,220
2010	111,534	67,946	32,008	0	0	99,954	(11,581)	(4,633)	(4,633)	402,516
2011	111,534	67,946	32,062	0	0	100,008	(11,526)	(4,611)	(4,611)	400,025
2012	111,534	67,946	32,008	0	0	99,954	(11,581)	(4,633)	(4,633)	397,707
2013	111,534	67,946	32,062	0	0	100,008	(11,526)	(4,611)	(4,611)	395,572
2014	111,534	67,946	32,008	0	0	99,954	(11,581)	(4,633)	(4,633)	393,585
2015	111,534	67,946	32,062	0	0	100,008	(11,526)	(4,611)	(4,611)	391,754
2016	111,534	67,946	32,008	0	0	99,954	(11,581)	(4,633)	(4,633)	390,050
2017	111,534	67,946	32,062	0	0	100,008	(11,526)	(4,611)	(4,611)	388,480
2018	111,534	67,946	16,004	0	0	83,950	(27,584)	(11,034)	(11,034)	385,002
2019	111,534	67,946	0	0	0	67,946	(43,588)	(17,435)	(17,435)	379,913
2020	111,534	67,946	0	0	0	67,946	(43,588)	(17,435)	(17,435)	375,201
2021	111,534	67,946	0	0	0	67,946	(43,588)	(17,435)	(17,435)	370,837
2022	111,534	67,946	0	0	0	67,946	(43,588)	(17,435)	(17,435)	366,798
2023	111,534	67,946	0	0	0	67,946	(43,588)	(17,435)	(17,435)	363,057
2024	111,534	67,946	0	0	0	67,946	(43,588)	(17,435)	(17,435)	359,593
2025	111,534	67,946	0	0	0	67,946	(43,588)	(17,435)	(17,435)	356,386
2026	111,534	67,946	0	0	0	67,946	(43,588)	(17,435)	(17,435)	353,417
2027	111,534	67,946	0	0	0	67,946	(43,588)	(17,435)	(17,435)	350,667
2028	111,534	67,946	0	0	0	67,946	(43,588)	(17,435)	(17,435)	348,122
2029	111,534	67,946	0	0	0	67,946	(43,588)	(17,435)	(17,435)	345,764
2030	111,534	67,946	0	0	0	67,946	(43,588)	(17,435)	(17,435)	343,582
2031	111,534	67,946	0	0	0	67,946	(43,588)	(17,435)	(17,435)	341,561
2032	111,534	67,946	0	0	0	67,946	(43,588)	(17,435)	(17,435)	339,689
2033	111,534	67,946	0	0	0	67,946	(43,588)	(17,435)	(17,435)	337,957
2034	111,534	67,946	0	0	0	67,946	(43,588)	(17,435)	(17,435)	336,353
2035	111,534	67,946	0	0	0	67,946	(43,588)	(17,435)	(17,435)	334,867
2036	111,534	67,946	0	0	0	67,946	(43,588)	(17,435)	(17,435)	333,492
2037	111,534	67,946	0	0	0	67,946	(43,588)	(17,435)	(17,435)	332,218
2038	111,534	67,946	0	0	0	67,946	(43,588)	(17,435)	(17,435)	331,039
2039	111,534	67,946	0	0	0	67,946	(43,588)	(17,435)	(17,435)	329,947
2040	111,534	67,946	0	0	0	67,946	(43,588)	(17,435)	(17,435)	328,936
2041	111,534	67,946	0	0	0	67,946	(43,588)	(17,435)	(17,435)	328,000
2042	23,273	14,178	0	0	0	14,178	(9,095)	(3,634)	(3,634)	**327,819**
Total	**$4,350,000**	**$2,650,000**	**$775,000**	**$0**	**$925,000**	**$4,350,000**	**$0**	**$0**	**$0**	
% of Total	100.00%	60.92%	17.82%	0.00%	21.26%	100.00%				

Figure 2.8

Credit: IRS Publication 946, "How to Depreciate Property."

Sample Retail Strip Plaza
Results of Cost Segregation Study
April 2000 Addition-Half Year

Cumulative present value on taxes deferred at 8%
$85,226
Total taxes deferred 2004-2007
$112,081
Total additional depreciation claimed 2004-2007
$280,202

Results of Cost Segregation Study	**Segregated**
Building - 39 Yr S/L	**$760,000**
Land Improve - 15 Yr 150% DB	**255,000**
Furnishings - 7 Yr 200% DB	**0**
Equipment - 5 Yr 200% DB	**235,000**
Total Recovery Property	**$1,250,000**

Assumptions	
Discount Rate	8%
Effective Fed'L Tax F	33.0%
Effective State Tax F	7.00%

Year	(A) Total Original Cost Deprec	Depreciation Based on Segregated Cost Recovery: 39 Yrs	15 Yrs	7 Yrs	5 Yrs	(B) Total Segregated Cost Deprec	(B)-(A) **Additional Depreciation Computed**	**Total Taxes Deferred**	Sub-total By Year	**Present Value of Cumulative Taxes Deferred**
2000	22,738	13,824	12,750	0	47,000	73,574	50,837	0		0
2001	32,050	19,486	24,225	0	75,200	118,911	86,861	0		0
2002	32,050	19,486	21,803	0	45,120	86,409	54,359	0		0
2003	32,050	19,486	19,635	0	27,072	66,193	34,143	0		0
Subtotal	**118,888**	**72,284**	**78,413**	**0**	**194,392**	**345,088**	**226,201**	**0**	**0**	**0**
										0
2004 Catch-Up							**226,201**	**90,480**		
2004	32,050	19,486	17,672	0	27,072	64,230	**32,180**	**12,872**	103,352	95,696
2005	32,050	19,486	15,887	0	13,536	48,909	**16,859**	**6,743**	6,743	101,477
2006	32,050	19,486	15,045	0	0	34,531	**2,481**	**993**	993	102,266
2007	32,050	19,486	15,045	0	0	34,531	**2,481**	**993**	$993	102,995
2008	32,050	19,486	15,071	0	0	34,557	2,507	1,002	$1,002	103,677
2009	32,050	19,486	15,045	0	0	34,531	2,481	993	$993	104,303
2010	32,050	19,486	15,071	0	0	34,557	2,507	1,002	$1,002	104,888
2011	32,050	19,486	15,045	0	0	34,531	2,481	993	$993	105,424
2012	32,050	19,486	15,071	0	0	34,557	2,507	1,002	$1,002	105,926
2013	32,050	19,486	15,045	0	0	34,531	2,481	993	$993	106,386
2014	32,050	19,486	15,071	0	0	34,557	2,507	1,002	$1,002	106,815
2015	32,050	19,486	7,523	0	0	27,009	(5,041)	(2,017)	($2,017)	106,014
2016	32,050	19,486	0	0	0	19,486	(12,564)	(5,025)	($5,025)	104,167
2017	32,050	19,486	0	0	0	19,486	(12,564)	(5,025)	($5,025)	102,456
2018	32,050	19,486	0	0	0	19,486	(12,564)	(5,025)	($5,025)	100,872
2019	32,050	19,486	0	0	0	19,486	(12,564)	(5,025)	($5,025)	99,405
2020	32,050	19,486	0	0	0	19,486	(12,564)	(5,025)	($5,025)	98,047
2021	32,050	19,486	0	0	0	19,486	(12,564)	(5,025)	($5,025)	96,789
2022	32,050	19,486	0	0	0	19,486	(12,564)	(5,025)	($5,025)	95,625
2023	32,050	19,486	0	0	0	19,486	(12,564)	(5,025)	($5,025)	94,547
2024	32,050	19,486	0	0	0	19,486	(12,564)	(5,025)	($5,025)	93,549
2025	32,050	19,486	0	0	0	19,486	(12,564)	(5,025)	($5,025)	92,624
2026	32,050	19,486	0	0	0	19,486	(12,564)	(5,025)	($5,025)	91,769
2027	32,050	19,486	0	0	0	19,486	(12,564)	(5,025)	($5,025)	90,976
2028	32,050	19,486	0	0	0	19,486	(12,564)	(5,025)	($5,025)	90,242
2029	32,050	19,486	0	0	0	19,486	(12,564)	(5,025)	($5,025)	89,563
2030	32,050	19,486	0	0	0	19,486	(12,564)	(5,025)	($5,025)	88,934
2031	32,050	19,486	0	0	0	19,486	(12,564)	(5,025)	($5,025)	88,351
2032	32,050	19,486	0	0	0	19,486	(12,564)	(5,025)	($5,025)	87,812
2033	32,050	19,486	0	0	0	19,486	(12,564)	(5,025)	($5,025)	87,313
2034	32,050	19,486	0	0	0	19,486	(12,564)	(5,025)	($5,025)	86,850
2035	32,050	19,486	0	0	0	19,486	(12,564)	(5,025)	($5,025)	86,422
2036	32,050	19,486	0	0	0	19,486	(12,564)	(5,025)	($5,025)	86,026
2037	32,050	19,486	0	0	0	19,486	(12,564)	(5,025)	($5,025)	85,659
2038	32,050	19,486	0	0	0	19,486	(12,564)	(5,025)	($5,025)	85,319
2039	9,363	5,692	0	0	0	5,692	(3,670)	(1,476)	($1,476)	**85,226**
Total	**$1,250,000**	**$760,000**	**$255,000**	**$0**	**$235,000**	**$1,250,000**	**$0**	**$0**	**$0**	
% of Total	100.00%	60.80%	20.40%	0.00%	18.80%	100.00%				

Figure 2.9
Credit: IRS Publication 946, "How to Depreciate Property."

Table A-1. **3-, 5-, 7-, 10-, 15-, and 20-Year Property Half-Year Convention**

Year	Depreciation rate for recovery period					
	3-year	5-year	7-year	10-year	15-year	20-year
1	33.33%	20.00%	14.29%	10.00%	5.00%	3.750%
2	44.45	32.00	24.49	18.00	9.50	7.219
3	14.81	19.20	17.49	14.40	8.55	6.677
4	7.41	11.52	12.49	11.52	7.70	6.177
5		11.52	8.93	9.22	6.93	5.713
6		5.76	8.92	7.37	6.23	5.285
7			8.93	6.55	5.90	4.888
8			4.46	6.55	5.90	4.522
9				6.56	5.91	4.462
10				6.55	5.90	4.461
11				3.28	5.91	4.462
12					5.90	4.461
13					5.91	4.462
14					5.90	4.461
15					5.91	4.462
16					2.95	4.461
17						4.462
18						4.461
19						4.462
20						4.461
21						2.231

Figure 2.10

Credit: IRS Publication 946, "How to Depreciate Property."

Table A-2. **3-, 5-, 7-, 10-, 15-, and 20-Year Property Mid-Quarter Convention Placed in Service in First Quarter**

Year	Depreciation rate for recovery period					
	3-year	5-year	7-year	10-year	15-year	20-year
1	58.33%	35.00%	25.00%	17.50%	8.75%	6.563%
2	27.78	26.00	21.43	16.50	9.13	7.000
3	12.35	15.60	15.31	13.20	8.21	6.482
4	1.54	11.01	10.93	10.56	7.39	5.996
5		11.01	8.75	8.45	6.65	5.546
6		1.38	8.74	6.76	5.99	5.130
7			8.75	6.55	5.90	4.746
8			1.09	6.55	5.91	4.459
9				6.56	5.90	4.459
10				6.55	5.91	4.459
11				0.82	5.90	4.459
12					5.91	4.460
13					5.90	4.459
14					5.91	4.460
15					5.90	4.459
16					0.74	4.460
17						4.459
18						4.460
19						4.459
20						4.460
21						0.557

Figure 2.11

Credit: IRS Publication 946, "How to Depreciate Property."

Table A-3. **3-, 5-, 7-, 10-, 15-, and 20-Year Property Mid-Quarter Convention Placed in Service in Second Quarter**

Year	Depreciation rate for recovery period					
	3-year	5-year	7-year	10-year	15-year	20-year
1	41.67%	25.00%	17.85%	12.50%	6.25%	4.688%
2	38.89	30.00	23.47	17.50	9.38	7.148
3	14.14	18.00	16.76	14.00	8.44	6.612
4	5.30	11.37	11.97	11.20	7.59	6.116
5		11.37	8.87	8.96	6.83	5.658
6		4.26	8.87	7.17	6.15	5.233
7			8.87	6.55	5.91	4.841
8			3.33	6.55	5.90	4.478
9				6.56	5.91	4.463
10				6.55	5.90	4.463
11				2.46	5.91	4.463
12					5.90	4.463
13					5.91	4.463
14					5.90	4.463
15					5.91	4.462
16					2.21	4.463
17						4.462
18						4.463
19						4.462
20						4.463
21						1.673

Figure 2.12

Credit: IRS Publication 946, "How to Depreciate Property."

Table A-4. **3-, 5-, 7-, 10-, 15-, and 20-Year Property Mid-Quarter Convention Placed in Service in Third Quarter**

Year	Depreciation rate for recovery period					
	3-year	5-year	7-year	10-year	15-year	20-year
1	25.00%	15.00%	10.71%	7.50%	3.75%	2.813%
2	50.00	34.00	25.51	18.50	9.63	7.289
3	16.67	20.40	18.22	14.80	8.66	6.742
4	8.33	12.24	13.02	11.84	7.80	6.237
5		11.30	9.30	9.47	7.02	5.769
6		7.06	8.85	7.58	6.31	5.336
7			8.86	6.55	5.90	4.936
8			5.53	6.55	5.90	4.566
9				6.56	5.91	4.460
10				6.55	5.90	4.460
11				4.10	5.91	4.460
12					5.90	4.460
13					5.91	4.461
14					5.90	4.460
15					5.91	4.461
16					3.69	4.460
17						4.461
18						4.460
19						4.461
20						4.460
21						2.788

Figure 2.13

Credit: IRS Publication 946, "How to Depreciate Property."

Table A-5. **3-, 5-, 7-, 10-, 15-, and 20-Year Property**
Mid-Quarter Convention
Placed in Service in Fourth Quarter

Year	Depreciation rate for recovery period					
	3-year	5-year	7-year	10-year	15-year	20-year
1	8.33%	5.00%	3.57%	2.50%	1.25%	0.938%
2	61.11	38.00	27.55	19.50	9.88	7.430
3	20.37	22.80	19.68	15.60	8.89	6.872
4	10.19	13.68	14.06	12.48	8.00	6.357
5		10.94	10.04	9.98	7.20	5.880
6		9.58	8.73	7.99	6.48	5.439
7			8.73	6.55	5.90	5.031
8			7.64	6.55	5.90	4.654
9				6.56	5.90	4.458
10				6.55	5.91	4.458
11				5.74	5.90	4.458
12					5.91	4.458
13					5.90	4.458
14					5.91	4.458
15					5.90	4.458
16					5.17	4.458
17						4.458
18						4.459
19						4.458
20						4.459
21						3.901

Figure 2.14
Credit: IRS Publication 946, "How to Depreciate Property."

Table A-6. **Residential Rental Property**
Mid-Month Convention
Straight Line—27.5 Years

Year	Month property placed in service											
	1	2	3	4	5	6	7	8	9	10	11	12
1	3.485%	3.182%	2.879%	2.576%	2.273%	1.970%	1.667%	1.364%	1.061%	0.758%	0.455%	0.152%
2–9	3.636	3.636	3.636	3.636	3.636	3.636	3.636	3.636	3.636	3.636	3.636	3.636
10	3.637	3.637	3.637	3.637	3.637	3.637	3.636	3.636	3.636	3.636	3.636	3.636
11	3.636	3.636	3.636	3.636	3.636	3.636	3.637	3.637	3.637	3.637	3.637	3.637
12	3.637	3.637	3.637	3.637	3.637	3.637	3.636	3.636	3.636	3.636	3.636	3.636
13	3.636	3.636	3.636	3.636	3.636	3.636	3.637	3.637	3.637	3.637	3.637	3.637
14	3.637	3.637	3.637	3.637	3.637	3.637	3.636	3.636	3.636	3.636	3.636	3.636
15	3.636	3.636	3.636	3.636	3.636	3.636	3.637	3.637	3.637	3.637	3.637	3.637
16	3.637	3.637	3.637	3.637	3.637	3.637	3.636	3.636	3.636	3.636	3.636	3.636
17	3.636	3.636	3.636	3.636	3.636	3.636	3.637	3.637	3.637	3.637	3.637	0.607
18	3.637	3.637	3.637	3.637	3.637	3.637	3.636	3.636	3.636	3.636	3.636	3.636
19	3.636	3.636	3.636	3.636	3.636	3.636	3.637	3.637	3.637	3.637	3.637	3.637
20	3.637	3.637	3.637	3.637	3.637	3.637	3.636	3.636	3.636	3.636	3.636	3.636
21	3.636	3.636	3.636	3.636	3.636	3.636	3.637	3.637	3.637	3.637	3.637	3.637
22	3.637	3.637	3.637	3.637	3.637	3.637	3.636	3.636	3.636	3.636	3.636	3.636
23	3.636	3.636	3.636	3.636	3.636	3.636	3.637	3.637	3.637	3.637	3.637	3.637
24	3.637	3.637	3.637	3.637	3.637	3.637	3.636	3.636	3.636	3.636	3.636	3.636
25	3.636	3.636	3.636	3.636	3.636	3.636	3.637	3.637	3.637	3.637	3.637	3.637
26	3.637	3.637	3.637	3.637	3.637	3.637	3.636	3.636	3.636	3.636	3.636	3.636
27	3.636	3.636	3.636	3.636	3.636	3.636	3.637	3.637	3.637	3.637	3.637	3.637
28	1.97	2.273	2.576	2.879	3.182	3.485	3.636	3.636	3.636	3.636	3.636	3.636
29							0.152	0.455	0.758	1.061	1.364	1.667

Figure 2.15
Credit: IRS Publication 946, "How to Depreciate Property."

Table A-7. **Nonresidential Real Property**
Mid-Month Convention
Straight Line—31.5 Years

Year	Month property placed in service											
	1	2	3	4	5	6	7	8	9	10	11	12
1	3.042%	2.778%	2.513%	2.249%	1.984%	1.720%	1.455%	1.190%	0.926%	0.661%	0.397%	0.132%
2–7	3.175	3.175	3.175	3.175	3.175	3.175	3.175	3.175	3.175	3.175	3.175	3.175
8	3.175	3.174	3.175	3.174	3.175	3.174	3.175	3.175	3.175	3.175	3.175	3.175
9	3.174	3.175	3.174	3.175	3.174	3.175	3.174	3.175	3.174	3.175	3.174	3.175
10	3.175	3.174	3.175	3.174	3.175	3.174	3.175	3.174	3.175	3.174	3.175	3.174
11	3.174	3.175	3.174	3.175	3.174	3.175	3.174	3.175	3.174	3.175	3.174	3.175
12	3.175	3.174	3.175	3.174	3.175	3.174	3.175	3.174	3.175	3.174	3.175	3.174
13	3.174	3.175	3.174	3.175	3.174	3.175	3.174	3.175	3.174	3.175	3.174	3.175
14	3.175	3.174	3.175	3.174	3.175	3.174	3.175	3.174	3.175	3.174	3.175	3.174
15	3.174	3.175	3.174	3.175	3.174	3.175	3.174	3.175	3.174	3.175	3.174	3.175
16	3.175	3.174	3.175	3.174	3.175	3.174	3.175	3.174	3.175	3.174	3.175	3.174
17	3.174	3.175	3.174	3.175	3.174	3.175	3.174	3.175	3.174	3.175	3.174	3.175
18	3.175	3.174	3.175	3.174	3.175	3.174	3.175	3.174	3.175	3.174	3.175	3.174
19	3.174	3.175	3.174	3.175	3.174	3.175	3.174	3.175	3.174	3.175	3.174	3.175
20	3.175	3.174	3.175	3.174	3.175	3.174	3.175	3.174	3.175	3.174	3.175	3.174
21	3.174	3.175	3.174	3.175	3.174	3.175	3.174	3.175	3.174	3.175	3.174	3.175
22	3.175	3.174	3.175	3.174	3.175	3.174	3.175	3.174	3.175	3.174	3.175	3.174
23	3.174	3.175	3.174	3.175	3.174	3.175	3.174	3.175	3.174	3.175	3.174	3.175
24	3.175	3.174	3.175	3.174	3.175	3.174	3.175	3.174	3.175	3.174	3.175	3.174
25	3.174	3.175	3.174	3.175	3.174	3.175	3.174	3.175	3.174	3.175	3.174	3.175
26	3.175	3.174	3.175	3.174	3.175	3.174	3.175	3.174	3.175	3.174	3.175	3.174
27	3.174	3.175	3.174	3.175	3.174	3.175	3.174	3.175	3.174	3.175	3.174	3.175
28	3.175	3.174	3.175	3.174	3.175	3.174	3.175	3.174	3.175	3.174	3.175	3.174
29	3.174	3.175	3.174	3.175	3.174	3.175	3.174	3.175	3.174	3.175	3.174	3.175
30	3.175	3.174	3.175	3.174	3.175	3.174	3.175	3.174	3.175	3.174	3.175	3.174
31	3.174	3.175	3.174	3.175	3.174	3.175	3.174	3.175	3.174	3.175	3.174	3.175
32	1.720	1.984	2.249	2.513	2.778	3.042	3.175	3.174	3.175	3.174	3.175	3.174
33							0.132	0.397	0.661	0.926	1.190	1.455

Figure 2.16
Credit: IRS Publication 946, "How to Depreciate Property."

Table A-7a. **Nonresidential Real Property**
Mid-Month Convention
Straight Line—39 Years

Year	Month property placed in service											
	1	2	3	4	5	6	7	8	9	10	11	12
1	2.461%	2.247%	2.033%	1.819%	1.605%	1.391%	1.177%	0.963%	0.749%	0.535%	0.321%	0.107%
2–39	2.564	2.564	2.564	2.564	2.564	2.564	2.564	2.564	2.564	2.564	2.564	2.564
40	0.107	0.321	0.535	0.749	0.963	1.177	1.391	1.605	1.819	2.033	2.247	2.461

Figure 2.17
Credit: IRS Publication 946, "How to Depreciate Property."

Table A-8. **Straight Line Method Half-Year Convention**

Year	Recovery periods in years												
	2.5	3	3.5	4	5	6	6.5	7	7.5	8	8.5	9	9.5
1	20.0%	16.67%	14.29%	12.5%	10.0%	8.33%	7.69%	7.14%	6.67%	6.25%	5.88%	5.56%	5.26%
2	40.0	33.33	28.57	25.0	20.0	16.67	15.39	14.29	13.33	12.50	11.77	11.11	10.53
3	40.0	33.33	28.57	25.0	20.0	16.67	15.38	14.29	13.33	12.50	11.76	11.11	10.53
4		16.67	28.57	25.0	20.0	16.67	15.39	14.28	13.33	12.50	11.77	11.11	10.53
5				12.5	20.0	16.66	15.38	14.29	13.34	12.50	11.76	11.11	10.52
6					10.0	16.67	15.39	14.28	13.33	12.50	11.77	11.11	10.53
7						8.33	15.38	14.29	13.34	12.50	11.76	11.11	10.52
8								7.14	13.33	12.50	11.77	11.11	10.53
9										6.25	11.76	11.11	10.52
10												5.56	10.53

Table A-8. *(Continued)*

Year	Recovery periods in years												
	10	10.5	11	11.5	12	12.5	13	13.5	14	15	16	16.5	17
1	5.0%	4.76%	4.55%	4.35%	4.17%	4.0%	3.85%	3.70%	3.57%	3.33%	3.13%	3.03%	2.94%
2	10.0	9.52	9.09	8.70	8.33	8.0	7.69	7.41	7.14	6.67	6.25	6.06	5.88
3	10.0	9.52	9.09	8.70	8.33	8.0	7.69	7.41	7.14	6.67	6.25	6.06	5.88
4	10.0	9.53	9.09	8.69	8.33	8.0	7.69	7.41	7.14	6.67	6.25	6.06	5.88
5	10.0	9.52	9.09	8.70	8.33	8.0	7.69	7.41	7.14	6.67	6.25	6.06	5.88
6	10.0	9.53	9.09	8.69	8.33	8.0	7.69	7.41	7.14	6.67	6.25	6.06	5.88
7	10.0	9.52	9.09	8.70	8.34	8.0	7.69	7.41	7.14	6.67	6.25	6.06	5.88
8	10.0	9.53	9.09	8.69	8.33	8.0	7.69	7.41	7.15	6.66	6.25	6.06	5.88
9	10.0	9.52	9.09	8.70	8.34	8.0	7.69	7.41	7.14	6.67	6.25	6.06	5.88
10	10.0	9.53	9.09	8.69	8.33	8.0	7.70	7.40	7.15	6.66	6.25	6.06	5.88
11	5.0	9.52	9.09	8.70	8.34	8.0	7.69	7.41	7.14	6.67	6.25	6.06	5.89
12			4.55	8.69	8.33	8.0	7.70	7.40	7.15	6.66	6.25	6.06	5.88
13					4.17	8.0	7.69	7.41	7.14	6.67	6.25	6.06	5.89
14							3.85	7.40	7.15	6.66	6.25	6.06	5.88
15									3.57	6.67	6.25	6.06	5.89
16										3.33	6.25	6.06	5.88
17											3.12	6.07	5.89
18													2.94

Figure 2.18

Credit: IRS Publication 946, "How to Depreciate Property."

Table A-8. *(Continued)*

Year	Recovery periods in years												
	18	19	20	22	24	25	26.5	28	30	35	40	45	50
1	2.78%	2.63%	2.5%	2.273%	2.083%	2.0%	1.887%	1.786%	1.667%	1.429%	1.25%	1.111%	1.0%
2	5.56	5.26	5.0	4.545	4.167	4.0	3.774	3.571	3.333	2.857	2.50	2.222	2.0
3	5.56	5.26	5.0	4.545	4.167	4.0	3.774	3.571	3.333	2.857	2.50	2.222	2.0
4	5.55	5.26	5.0	4.545	4.167	4.0	3.774	3.571	3.333	2.857	2.50	2.222	2.0
5	5.56	5.26	5.0	4.546	4.167	4.0	3.774	3.571	3.333	2.857	2.50	2.222	2.0
6	5.55	5.26	5.0	4.545	4.167	4.0	3.774	3.571	3.333	2.857	2.50	2.222	2.0
7	5.56	5.26	5.0	4.546	4.167	4.0	3.773	3.572	3.333	2.857	2.50	2.222	2.0
8	5.55	5.26	5.0	4.545	4.167	4.0	3.774	3.571	3.333	2.857	2.50	2.222	2.0
9	5.56	5.27	5.0	4.546	4.167	4.0	3.773	3.572	3.333	2.857	2.50	2.222	2.0
10	5.55	5.26	5.0	4.545	4.167	4.0	3.774	3.571	3.333	2.857	2.50	2.222	2.0
11	5.56	5.27	5.0	4.546	4.166	4.0	3.773	3.572	3.333	2.857	2.50	2.222	2.0
12	5.55	5.26	5.0	4.545	4.167	4.0	3.774	3.571	3.333	2.857	2.50	2.222	2.0
13	5.56	5.27	5.0	4.546	4.166	4.0	3.773	3.572	3.334	2.857	2.50	2.222	2.0
14	5.55	5.26	5.0	4.545	4.167	4.0	3.773	3.571	3.333	2.857	2.50	2.222	2.0
15	5.56	5.27	5.0	4.546	4.166	4.0	3.774	3.572	3.334	2.857	2.50	2.222	2.0
16	5.55	5.26	5.0	4.545	4.167	4.0	3.773	3.571	3.333	2.857	2.50	2.222	2.0
17	5.56	5.27	5.0	4.546	4.166	4.0	3.774	3.572	3.334	2.857	2.50	2.222	2.0
18	5.55	5.26	5.0	4.545	4.167	4.0	3.773	3.571	3.333	2.857	2.50	2.222	2.0
19	2.78	5.27	5.0	4.546	4.166	4.0	3.774	3.572	3.334	2.857	2.50	2.222	2.0
20		2.63	5.0	4.545	4.167	4.0	3.773	3.571	3.333	2.857	2.50	2.222	2.0
21			2.5	4.546	4.166	4.0	3.774	3.572	3.334	2.857	2.50	2.222	2.0
22				4.545	4.167	4.0	3.773	3.571	3.333	2.857	2.50	2.222	2.0
23				2.273	4.166	4.0	3.774	3.572	3.334	2.857	2.50	2.222	2.0
24					4.167	4.0	3.773	3.571	3.333	2.857	2.50	2.222	2.0
25					2.083	4.0	3.774	3.572	3.334	2.857	2.50	2.222	2.0
26						2.0	3.773	3.571	3.333	2.857	2.50	2.222	2.0
27							3.774	3.572	3.334	2.857	2.50	2.223	2.0
28								3.571	3.333	2.858	2.50	2.222	2.0
29								1.786	3.334	2.857	2.50	2.223	2.0
30									3.333	2.858	2.50	2.222	2.0
31									1.667	2.857	2.50	2.223	2.0
32										2.858	2.50	2.222	2.0
33										2.857	2.50	2.223	2.0
34										2.858	2.50	2.222	2.0
35										2.857	2.50	2.223	2.0
36										1.429	2.50	2.222	2.0
37											2.50	2.223	2.0
38											2.50	2.222	2.0
39											2.50	2.223	2.0
40											2.50	2.222	2.0
41											1.25	2.223	2.0
42												2.222	2.0
43												2.223	2.0
44												2.222	2.0
45												2.223	2.0
46												1.111	2.0
47–50													2.0
51													1.0

Figure 2.18 continued

Credit: IRS Publication 946, "How to Depreciate Property."

Table A-9. **Straight Line Method**
Mid-Quarter Convention
Placed in Service in First Quarter

Year	Recovery periods in years												
	2.5	**3**	**3.5**	**4**	**5**	**6**	**6.5**	**7**	**7.5**	**8**	**8.5**	**9**	**9.5**
1	35.0%	29.17%	25.00%	21.88%	17.5%	14.58%	13.46%	12.50%	11.67%	10.94%	10.29%	9.72%	9.21%
2	40.0	33.33	28.57	25.00	20.0	16.67	15.38	14.29	13.33	12.50	11.77	11.11	10.53
3	25.0	33.33	28.57	25.00	20.0	16.67	15.39	14.28	13.33	12.50	11.76	11.11	10.53
4		4.17	17.86	25.00	20.0	16.67	15.38	14.29	13.33	12.50	11.77	11.11	10.53
5				3.12	20.0	16.66	15.39	14.28	13.34	12.50	11.76	11.11	10.52
6					2.5	16.67	15.38	14.29	13.33	12.50	11.77	11.11	10.53
7						2.08	9.62	14.28	13.34	12.50	11.76	11.11	10.52
8								1.79	8.33	12.50	11.77	11.12	10.53
9										1.56	7.35	11.11	10.52
10												1.39	6.58

Table A-9. *(Continued)*

Year	Recovery periods in years												
	10	**10.5**	**11**	**11.5**	**12**	**12.5**	**13**	**13.5**	**14**	**15**	**16**	**16.5**	**17**
1	8.75%	8.33%	7.95%	7.61%	7.29%	7.0%	6.73%	6.48%	6.25%	5.83%	5.47%	5.30%	5.15%
2	10.00	9.52	9.09	8.70	8.33	8.0	7.69	7.41	7.14	6.67	6.25	6.06	5.88
3	10.00	9.52	9.09	8.70	8.33	8.0	7.69	7.41	7.14	6.67	6.25	6.06	5.88
4	10.00	9.53	9.09	8.69	8.33	8.0	7.69	7.41	7.14	6.67	6.25	6.06	5.88
5	10.00	9.52	9.09	8.70	8.33	8.0	7.69	7.41	7.14	6.67	6.25	6.06	5.88
6	10.00	9.53	9.09	8.69	8.34	8.0	7.69	7.41	7.14	6.67	6.25	6.06	5.88
7	10.00	9.52	9.09	8.70	8.33	8.0	7.69	7.41	7.14	6.67	6.25	6.06	5.88
8	10.00	9.53	9.09	8.69	8.34	8.0	7.69	7.41	7.15	6.66	6.25	6.06	5.88
9	10.00	9.52	9.09	8.70	8.33	8.0	7.70	7.40	7.14	6.67	6.25	6.06	5.88
10	10.00	9.53	9.10	8.69	8.34	8.0	7.69	7.41	7.15	6.66	6.25	6.06	5.88
11	1.25	5.95	9.09	8.70	8.33	8.0	7.70	7.40	7.14	6.67	6.25	6.06	5.88
12			1.14	5.43	8.34	8.0	7.69	7.41	7.15	6.66	6.25	6.06	5.89
13					1.04	5.0	7.70	7.40	7.14	6.67	6.25	6.06	5.88
14							0.96	4.63	7.15	6.66	6.25	6.06	5.89
15									0.89	6.67	6.25	6.06	5.88
16										0.83	6.25	6.07	5.89
17											0.78	3.79	5.88
18													0.74

Figure 2.19
Credit: IRS Publication 946, "How to Depreciate Property."

Table A-9. *(Continued)*

Year	Recovery periods in years												
	18	**19**	**20**	**22**	**24**	**25**	**26.5**	**28**	**30**	**35**	**40**	**45**	**50**
1	4.86%	4.61%	4.375%	3.977%	3.646%	3.5%	3.302%	3.125%	2.917%	2.500%	2.188%	1.944%	1.75%
2	5.56	5.26	5.000	4.545	4.167	4.0	3.774	3.571	3.333	2.857	2.500	2.222	2.00
3	5.56	5.26	5.000	4.545	4.167	4.0	3.774	3.571	3.333	2.857	2.500	2.222	2.00
4	5.56	5.26	5.000	4.546	4.167	4.0	3.774	3.571	3.333	2.857	2.500	2.222	2.00
5	5.55	5.26	5.000	4.545	4.167	4.0	3.774	3.571	3.333	2.857	2.500	2.222	2.00
6	5.56	5.26	5.000	4.546	4.167	4.0	3.774	3.572	3.333	2.857	2.500	2.222	2.00
7	5.55	5.26	5.000	4.545	4.167	4.0	3.773	3.571	3.333	2.857	2.500	2.222	2.00
8	5.56	5.26	5.000	4.546	4.167	4.0	3.774	3.572	3.333	2.857	2.500	2.222	2.00
9	5.55	5.26	5.000	4.545	4.167	4.0	3.773	3.571	3.333	2.857	2.500	2.222	2.00
10	5.56	5.27	5.000	4.546	4.166	4.0	3.774	3.572	3.333	2.857	2.500	2.222	2.00
11	5.55	5.26	5.000	4.545	4.167	4.0	3.773	3.571	3.333	2.857	2.500	2.222	2.00
12	5.56	5.27	5.000	4.546	4.166	4.0	3.774	3.572	3.333	2.857	2.500	2.222	2.00
13	5.55	5.26	5.000	4.545	4.167	4.0	3.773	3.571	3.334	2.857	2.500	2.222	2.00
14	5.56	5.27	5.000	4.546	4.166	4.0	3.774	3.572	3.333	2.857	2.500	2.222	2.00
15	5.55	5.26	5.000	4.545	4.167	4.0	3.773	3.571	3.334	2.857	2.500	2.222	2.00
16	5.56	5.27	5.000	4.546	4.166	4.0	3.774	3.572	3.333	2.857	2.500	2.222	2.00
17	5.55	5.26	5.000	4.545	4.167	4.0	3.773	3.571	3.334	2.857	2.500	2.222	2.00
18	5.56	5.27	5.000	4.546	4.166	4.0	3.774	3.572	3.333	2.857	2.500	2.222	2.00
19	0.69	5.26	5.000	4.545	4.167	4.0	3.773	3.571	3.334	2.857	2.500	2.222	2.00
20		0.66	5.000	4.546	4.166	4.0	3.774	3.572	3.333	2.857	2.500	2.222	2.00
21			0.625	4.545	4.167	4.0	3.773	3.571	3.334	2.857	2.500	2.222	2.00
22				4.546	4.166	4.0	3.774	3.572	3.333	2.857	2.500	2.222	2.00
23				0.568	4.167	4.0	3.773	3.571	3.334	2.857	2.500	2.222	2.00
24					4.166	4.0	3.774	3.572	3.333	2.857	2.500	2.222	2.00
25					0.521	4.0	3.773	3.571	3.334	2.857	2.500	2.222	2.00
26						0.5	3.774	3.572	3.333	2.857	2.500	2.223	2.00
27							2.358	3.571	3.334	2.858	2.500	2.222	2.00
28								3.572	3.333	2.857	2.500	2.223	2.00
29								0.446	3.334	2.858	2.500	2.222	2.00
30									3.333	2.857	2.500	2.223	2.00
31									0.417	2.858	2.500	2.222	2.00
32										2.857	2.500	2.223	2.00
33										2.858	2.500	2.222	2.00
34										2.857	2.500	2.223	2.00
35										2.858	2.500	2.222	2.00
36										0.357	2.500	2.223	2.00
37											2.500	2.222	2.00
38											2.500	2.223	2.00
39											2.500	2.222	2.00
40											2.500	2.223	2.00
41											0.312	2.222	2.00
42												2.223	2.00
43												2.222	2.00
44												2.223	2.00
45												2.222	2.00
46												0.278	2.00
47–50													2.00
51													0.25

Figure 2.19 continued

Credit: IRS Publication 946, "How to Depreciate Property."

Table A-10. **Straight Line Method**
Mid-Quarter Convention
Placed in Service in Second Quarter

Year	Recovery periods in years												
	2.5	**3**	**3.5**	**4**	**5**	**6**	**6.5**	**7**	**7.5**	**8**	**8.5**	**9**	**9.5**
1	25.0%	20.83%	17.86%	15.63%	12.5%	10.42%	9.62%	8.93%	8.33%	7.81%	7.35%	6.94%	6.58%
2	40.0	33.33	28.57	25.00	20.0	16.67	15.38	14.29	13.33	12.50	11.77	11.11	10.53
3	35.0	33.34	28.57	25.00	20.0	16.67	15.38	14.28	13.33	12.50	11.76	11.11	10.53
4		12.50	25.00	25.00	20.0	16.66	15.39	14.29	13.34	12.50	11.77	11.11	10.53
5				9.37	20.0	16.67	15.38	14.28	13.33	12.50	11.76	11.11	10.52
6					7.5	16.66	15.39	14.29	13.34	12.50	11.77	11.11	10.53
7						6.25	13.46	14.28	13.33	12.50	11.76	11.11	10.52
8								5.36	11.67	12.50	11.77	11.12	10.53
9										4.69	10.29	11.11	10.52
10												4.17	9.21

Table A-10. *(Continued)*

Year	Recovery periods in years												
	10	**10.5**	**11**	**11.5**	**12**	**12.5**	**13**	**13.5**	**14**	**15**	**16**	**16.5**	**17**
1	6.25%	5.95%	5.68%	5.43%	5.21%	5.0%	4.81%	4.63%	4.46%	4.17%	3.91%	3.79%	3.68%
2	10.00	9.52	9.09	8.70	8.33	8.0	7.69	7.41	7.14	6.67	6.25	6.06	5.88
3	10.00	9.52	9.09	8.70	8.33	8.0	7.69	7.41	7.14	6.67	6.25	6.06	5.88
4	10.00	9.53	9.09	8.70	8.33	8.0	7.69	7.41	7.14	6.67	6.25	6.06	5.88
5	10.00	9.52	9.09	8.69	8.33	8.0	7.69	7.41	7.14	6.67	6.25	6.06	5.88
6	10.00	9.53	9.09	8.70	8.33	8.0	7.69	7.41	7.14	6.67	6.25	6.06	5.88
7	10.00	9.52	9.09	8.69	8.34	8.0	7.69	7.41	7.15	6.66	6.25	6.06	5.88
8	10.00	9.53	9.09	8.70	8.33	8.0	7.69	7.41	7.14	6.67	6.25	6.06	5.88
9	10.00	9.52	9.09	8.69	8.34	8.0	7.69	7.40	7.15	6.66	6.25	6.06	5.88
10	10.00	9.53	9.09	8.70	8.33	8.0	7.70	7.41	7.14	6.67	6.25	6.06	5.88
11	3.75	8.33	9.10	8.69	8.34	8.0	7.69	7.40	7.15	6.66	6.25	6.06	5.88
12			3.41	7.61	8.33	8.0	7.70	7.41	7.14	6.67	6.25	6.06	5.89
13					3.13	7.0	7.69	7.40	7.15	6.66	6.25	6.06	5.88
14							2.89	6.48	7.14	6.67	6.25	6.06	5.89
15									2.68	6.66	6.25	6.06	5.88
16										2.50	6.25	6.06	5.89
17											2.34	5.31	5.88
18													2.21

Figure 2.20

Credit: IRS Publication 946, "How to Depreciate Property."

Table A-10. *(Continued)*

Year	Recovery periods in years												
	18	**19**	**20**	**22**	**24**	**25**	**26.5**	**28**	**30**	**35**	**40**	**45**	**50**
1	3.47%	3.29%	3.125%	2.841%	2.604%	2.5%	2.358%	2.232%	2.083%	1.786%	1.563%	1.389%	1.25%
2	5.56	5.26	5.000	4.545	4.167	4.0	3.774	3.571	3.333	2.857	2.500	2.222	2.00
3	5.56	5.26	5.000	4.545	4.167	4.0	3.774	3.571	3.333	2.857	2.500	2.222	2.00
4	5.56	5.26	5.000	4.545	4.167	4.0	3.774	3.571	3.333	2.857	2.500	2.222	2.00
5	5.55	5.26	5.000	4.546	4.167	4.0	3.774	3.571	3.333	2.857	2.500	2.222	2.00
6	5.56	5.26	5.000	4.545	4.167	4.0	3.774	3.572	3.333	2.857	2.500	2.222	2.00
7	5.55	5.26	5.000	4.546	4.167	4.0	3.774	3.571	3.333	2.857	2.500	2.222	2.00
8	5.56	5.26	5.000	4.545	4.167	4.0	3.773	3.572	3.333	2.857	2.500	2.222	2.00
9	5.55	5.27	5.000	4.546	4.167	4.0	3.774	3.571	3.333	2.857	2.500	2.222	2.00
10	5.56	5.26	5.000	4.545	4.167	4.0	3.773	3.572	3.333	2.857	2.500	2.222	2.00
11	5.55	5.27	5.000	4.546	4.166	4.0	3.774	3.571	3.333	2.857	2.500	2.222	2.00
12	5.56	5.26	5.000	4.545	4.167	4.0	3.773	3.572	3.334	2.857	2.500	2.222	2.00
13	5.55	5.27	5.000	4.546	4.166	4.0	3.774	3.571	3.333	2.857	2.500	2.222	2.00
14	5.56	5.26	5.000	4.545	4.167	4.0	3.773	3.572	3.334	2.857	2.500	2.222	2.00
15	5.55	5.27	5.000	4.546	4.166	4.0	3.774	3.571	3.333	2.857	2.500	2.222	2.00
16	5.56	5.26	5.000	4.545	4.167	4.0	3.773	3.572	3.334	2.857	2.500	2.222	2.00
17	5.55	5.27	5.000	4.546	4.166	4.0	3.774	3.571	3.333	2.857	2.500	2.222	2.00
18	5.56	5.26	5.000	4.545	4.167	4.0	3.773	3.572	3.334	2.857	2.500	2.222	2.00
19	2.08	5.27	5.000	4.546	4.166	4.0	3.774	3.571	3.333	2.857	2.500	2.222	2.00
20		1.97	5.000	4.545	4.167	4.0	3.773	3.572	3.334	2.857	2.500	2.222	2.00
21			1.875	4.546	4.166	4.0	3.774	3.571	3.333	2.857	2.500	2.222	2.00
22				4.545	4.167	4.0	3.773	3.572	3.334	2.857	2.500	2.222	2.00
23				1.705	4.166	4.0	3.774	3.571	3.333	2.857	2.500	2.222	2.00
24					4.167	4.0	3.773	3.572	3.334	2.857	2.500	2.222	2.00
25					1.562	4.0	3.774	3.571	3.333	2.857	2.500	2.222	2.00
26						1.5	3.773	3.572	3.334	2.857	2.500	2.222	2.00
27							3.302	3.571	3.333	2.857	2.500	2.223	2.00
28								3.572	3.334	2.858	2.500	2.222	2.00
29								1.339	3.333	2.857	2.500	2.223	2.00
30									3.334	2.858	2.500	2.222	2.00
31									1.250	2.857	2.500	2.223	2.00
32										2.858	2.500	2.222	2.00
33										2.857	2.500	2.223	2.00
34										2.858	2.500	2.222	2.00
35										2.857	2.500	2.223	2.00
36										1.072	2.500	2.222	2.00
37											2.500	2.223	2.00
38											2.500	2.222	2.00
39											2.500	2.223	2.00
40											2.500	2.222	2.00
41											0.937	2.223	2.00
42												2.222	2.00
43												2.223	2.00
44												2.222	2.00
45												2.223	2.00
46												0.833	2.00
47–50													2.00
51													0.75

Figure 2.20 continued

Credit: IRS Publication 946, "How to Depreciate Property."

Table A-11. **Straight Line Method**
Mid-Quarter Convention
Placed in Service in Third Quarter

Year	Recovery periods in years												
	2.5	**3**	**3.5**	**4**	**5**	**6**	**6.5**	**7**	**7.5**	**8**	**8.5**	**9**	**9.5**
1	15.0%	12.50%	10.71%	9.38%	7.5%	6.25%	5.77%	5.36%	5.00%	4.69%	4.41%	4.17%	3.95%
2	40.0	33.33	28.57	25.00	20.0	16.67	15.38	14.29	13.33	12.50	11.76	11.11	10.53
3	40.0	33.34	28.57	25.00	20.0	16.67	15.39	14.28	13.33	12.50	11.77	11.11	10.53
4	5.0	20.83	28.58	25.00	20.0	16.66	15.38	14.29	13.33	12.50	11.76	11.11	10.52
5			3.57	15.62	20.0	16.67	15.39	14.28	13.34	12.50	11.77	11.11	10.53
6					12.5	16.66	15.38	14.29	13.33	12.50	11.76	11.11	10.52
7						10.42	15.39	14.28	13.34	12.50	11.77	11.11	10.53
8							1.92	8.93	13.33	12.50	11.76	11.11	10.52
9									1.67	7.81	11.77	11.11	10.53
10											1.47	6.95	10.52
11													1.32

Table A-11. *(Continued)*

Year	Recovery periods in years												
	10	**10.5**	**11**	**11.5**	**12**	**12.5**	**13**	**13.5**	**14**	**15**	**16**	**16.5**	**17**
1	3.75%	3.57%	3.41%	3.26%	3.13%	3.0%	2.88%	2.78%	2.68%	2.50%	2.34%	2.27%	2.21%
2	10.00	9.52	9.09	8.70	8.33	8.0	7.69	7.41	7.14	6.67	6.25	6.06	5.88
3	10.00	9.52	9.09	8.70	8.33	8.0	7.69	7.41	7.14	6.67	6.25	6.06	5.88
4	10.00	9.52	9.09	8.69	8.33	8.0	7.69	7.41	7.14	6.67	6.25	6.06	5.88
5	10.00	9.53	9.09	8.70	8.33	8.0	7.69	7.41	7.14	6.67	6.25	6.06	5.88
6	10.00	9.52	9.09	8.69	8.33	8.0	7.69	7.41	7.14	6.67	6.25	6.06	5.88
7	10.00	9.53	9.09	8.70	8.34	8.0	7.69	7.41	7.14	6.66	6.25	6.06	5.88
8	10.00	9.52	9.09	8.69	8.33	8.0	7.70	7.40	7.14	6.67	6.25	6.06	5.88
9	10.00	9.53	9.09	8.70	8.34	8.0	7.69	7.41	7.15	6.66	6.25	6.06	5.88
10	10.00	9.52	9.09	8.69	8.33	8.0	7.70	7.40	7.14	6.67	6.25	6.06	5.88
11	6.25	9.53	9.10	8.70	8.34	8.0	7.69	7.41	7.15	6.66	6.25	6.06	5.88
12		1.19	5.68	8.69	8.33	8.0	7.70	7.40	7.14	6.67	6.25	6.06	5.89
13				1.09	5.21	8.0	7.69	7.41	7.15	6.66	6.25	6.06	5.88
14						1.0	4.81	7.40	7.14	6.67	6.25	6.06	5.89
15								0.93	4.47	6.66	6.25	6.06	5.88
16										4.17	6.25	6.07	5.89
17											3.91	6.06	5.88
18												0.76	3.68

Figure 2.21
Credit: IRS Publication 946, "How to Depreciate Property."

Table A-11. *(Continued)*

Year	Recovery periods in years												
	18	19	20	22	24	25	26.5	28	30	35	40	45	50
1	2.08%	1.97%	1.875%	1.705%	1.563%	1.5%	1.415%	1.339%	1.250%	1.071%	0.938%	0.833%	0.75%
2	5.56	5.26	5.000	4.545	4.167	4.0	3.774	3.571	3.333	2.857	2.500	2.222	2.00
3	5.56	5.26	5.000	4.545	4.167	4.0	3.774	3.571	3.333	2.857	2.500	2.222	2.00
4	5.56	5.26	5.000	4.545	4.167	4.0	3.774	3.571	3.333	2.857	2.500	2.222	2.00
5	5.55	5.26	5.000	4.546	4.167	4.0	3.774	3.571	3.333	2.857	2.500	2.222	2.00
6	5.56	5.26	5.000	4.545	4.167	4.0	3.774	3.572	3.333	2.857	2.500	2.222	2.00
7	5.55	5.26	5.000	4.546	4.167	4.0	3.773	3.571	3.333	2.857	2.500	2.222	2.00
8	5.56	5.26	5.000	4.545	4.167	4.0	3.774	3.572	3.333	2.857	2.500	2.222	2.00
9	5.55	5.27	5.000	4.546	4.166	4.0	3.773	3.571	3.333	2.857	2.500	2.222	2.00
10	5.56	5.26	5.000	4.545	4.167	4.0	3.774	3.572	3.333	2.857	2.500	2.222	2.00
11	5.55	5.27	5.000	4.546	4.166	4.0	3.773	3.571	3.333	2.857	2.500	2.222	2.00
12	5.56	5.26	5.000	4.545	4.167	4.0	3.774	3.572	3.334	2.857	2.500	2.222	2.00
13	5.55	5.27	5.000	4.546	4.166	4.0	3.773	3.571	3.333	2.857	2.500	2.222	2.00
14	5.56	5.26	5.000	4.545	4.167	4.0	3.774	3.572	3.334	2.857	2.500	2.222	2.00
15	5.55	5.27	5.000	4.546	4.166	4.0	3.773	3.571	3.333	2.857	2.500	2.222	2.00
16	5.56	5.26	5.000	4.545	4.167	4.0	3.774	3.572	3.334	2.857	2.500	2.222	2.00
17	5.55	5.27	5.000	4.546	4.166	4.0	3.773	3.571	3.333	2.857	2.500	2.222	2.00
18	5.56	5.26	5.000	4.545	4.167	4.0	3.774	3.572	3.334	2.857	2.500	2.222	2.00
19	3.47	5.27	5.000	4.546	4.166	4.0	3.773	3.571	3.333	2.857	2.500	2.222	2.00
20		3.29	5.000	4.545	4.167	4.0	3.774	3.572	3.334	2.857	2.500	2.222	2.00
21			3.125	4.546	4.166	4.0	3.773	3.571	3.333	2.857	2.500	2.222	2.00
22				4.545	4.167	4.0	3.774	3.572	3.334	2.857	2.500	2.222	2.00
23				2.841	4.166	4.0	3.773	3.571	3.333	2.857	2.500	2.222	2.00
24					4.167	4.0	3.774	3.572	3.334	2.857	2.500	2.222	2.00
25					2.604	4.0	3.773	3.571	3.333	2.857	2.500	2.222	2.00
26						2.5	3.774	3.572	3.334	2.858	2.500	2.222	2.00
27							3.773	3.571	3.333	2.857	2.500	2.223	2.00
28							0.472	3.572	3.334	2.858	2.500	2.222	2.00
29								2.232	3.333	2.857	2.500	2.223	2.00
30									3.334	2.858	2.500	2.222	2.00
31									2.083	2.857	2.500	2.223	2.00
32										2.858	2.500	2.222	2.00
33										2.857	2.500	2.223	2.00
34										2.858	2.500	2.222	2.00
35										2.857	2.500	2.223	2.00
36										1.786	2.500	2.222	2.00
37											2.500	2.223	2.00
38											2.500	2.222	2.00
39											2.500	2.223	2.00
40											2.500	2.222	2.00
41											1.562	2.223	2.00
42												2.222	2.00
43												2.223	2.00
44												2.222	2.00
45												2.223	2.00
46												1.389	2.00
47–50													2.00
51													1.25

Figure 2.21 continued

Credit: IRS Publication 946, "How to Depreciate Property."

Table A-12. **Straight Line Method**
Mid-Quarter Convention
Placed in Service in Fourth Quarter

Year	Recovery periods in years												
	2.5	**3**	**3.5**	**4**	**5**	**6**	**6.5**	**7**	**7.5**	**8**	**8.5**	**9**	**9.5**
1	5.0%	4.17%	3.57%	3.13%	2.5%	2.08%	1.92%	1.79%	1.67%	1.56%	1.47%	1.39%	1.32%
2	40.0	33.33	28.57	25.00	20.0	16.67	15.39	14.29	13.33	12.50	11.76	11.11	10.53
3	40.0	33.33	28.57	25.00	20.0	16.67	15.38	14.28	13.33	12.50	11.77	11.11	10.53
4	15.0	29.17	28.57	25.00	20.0	16.67	15.39	14.29	13.33	12.50	11.76	11.11	10.52
5			10.72	21.87	20.0	16.66	15.38	14.28	13.33	12.50	11.77	11.11	10.53
6					17.5	16.67	15.39	14.29	13.34	12.50	11.76	11.11	10.52
7						14.58	15.38	14.28	13.33	12.50	11.77	11.11	10.53
8							5.77	12.50	13.34	12.50	11.76	11.11	10.52
9									5.00	10.94	11.77	11.11	10.53
10											4.41	9.73	10.52
11													3.95

Table A-12. *(Continued)*

Year	Recovery periods in years												
	10	**10.5**	**11**	**11.5**	**12**	**12.5**	**13**	**13.5**	**14**	**15**	**16**	**16.5**	**17**
1	1.25%	1.19%	1.14%	1.09%	1.04%	1.0%	0.96%	0.93%	0.89%	0.83%	0.78%	0.76%	0.74%
2	10.00	9.52	9.09	8.70	8.33	8.0	7.69	7.41	7.14	6.67	6.25	6.06	5.88
3	10.00	9.52	9.09	8.69	8.33	8.0	7.69	7.41	7.14	6.67	6.25	6.06	5.88
4	10.00	9.52	9.09	8.70	8.33	8.0	7.69	7.41	7.14	6.67	6.25	6.06	5.88
5	10.00	9.53	9.09	8.69	8.33	8.0	7.69	7.41	7.14	6.67	6.25	6.06	5.88
6	10.00	9.52	9.09	8.70	8.34	8.0	7.69	7.41	7.14	6.67	6.25	6.06	5.88
7	10.00	9.53	9.09	8.69	8.33	8.0	7.69	7.41	7.14	6.67	6.25	6.06	5.88
8	10.00	9.52	9.09	8.70	8.34	8.0	7.69	7.40	7.15	6.66	6.25	6.06	5.88
9	10.00	9.53	9.09	8.69	8.33	8.0	7.70	7.41	7.14	6.67	6.25	6.06	5.88
10	10.00	9.52	9.09	8.70	8.34	8.0	7.69	7.40	7.15	6.66	6.25	6.06	5.88
11	8.75	9.53	9.09	8.69	8.33	8.0	7.70	7.41	7.14	6.67	6.25	6.06	5.88
12		3.57	7.96	8.70	8.34	8.0	7.69	7.40	7.15	6.66	6.25	6.06	5.89
13				3.26	7.29	8.0	7.70	7.41	7.14	6.67	6.25	6.06	5.88
14						3.0	6.73	7.40	7.15	6.66	6.25	6.06	5.89
15								2.78	6.25	6.67	6.25	6.06	5.88
16										5.83	6.25	6.06	5.89
17											5.47	6.07	5.88
18												2.27	5.15

Figure 2.22
Credit: IRS Publication 946, "How to Depreciate Property."

Table A-12. *(Continued)*

Year	Recovery periods in years												
	18	19	20	22	24	25	26.5	28	30	35	40	45	50
1	0.69%	0.66%	0.625%	0.568%	0.521%	0.5%	0.472%	0.446%	0.417%	0.357%	0.313%	0.278%	0.25%
2	5.56	5.26	5.000	4.545	4.167	4.0	3.774	3.571	3.333	2.857	2.500	2.222	2.00
3	5.56	5.26	5.000	4.545	4.167	4.0	3.774	3.571	3.333	2.857	2.500	2.222	2.00
4	5.56	5.26	5.000	4.546	4.167	4.0	3.774	3.571	3.333	2.857	2.500	2.222	2.00
5	5.55	5.26	5.000	4.545	4.167	4.0	3.774	3.571	3.333	2.857	2.500	2.222	2.00
6	5.56	5.26	5.000	4.546	4.167	4.0	3.773	3.572	3.333	2.857	2.500	2.222	2.00
7	5.55	5.26	5.000	4.545	4.167	4.0	3.774	3.571	3.333	2.857	2.500	2.222	2.00
8	5.56	5.26	5.000	4.546	4.167	4.0	3.773	3.572	3.333	2.857	2.500	2.222	2.00
9	5.55	5.26	5.000	4.545	4.167	4.0	3.774	3.571	3.333	2.857	2.500	2.222	2.00
10	5.56	5.27	5.000	4.546	4.166	4.0	3.773	3.572	3.333	2.857	2.500	2.222	2.00
11	5.55	5.26	5.000	4.545	4.167	4.0	3.774	3.571	3.333	2.857	2.500	2.222	2.00
12	5.56	5.27	5.000	4.546	4.166	4.0	3.773	3.572	3.333	2.857	2.500	2.222	2.00
13	5.55	5.26	5.000	4.545	4.167	4.0	3.774	3.571	3.334	2.857	2.500	2.222	2.00
14	5.56	5.27	5.000	4.546	4.166	4.0	3.773	3.572	3.333	2.857	2.500	2.222	2.00
15	5.55	5.26	5.000	4.545	4.167	4.0	3.774	3.571	3.334	2.857	2.500	2.222	2.00
16	5.56	5.27	5.000	4.546	4.166	4.0	3.773	3.572	3.333	2.857	2.500	2.222	2.00
17	5.55	5.26	5.000	4.545	4.167	4.0	3.774	3.571	3.334	2.857	2.500	2.222	2.00
18	5.56	5.27	5.000	4.546	4.166	4.0	3.773	3.572	3.333	2.857	2.500	2.222	2.00
19	4.86	5.26	5.000	4.545	4.167	4.0	3.774	3.571	3.334	2.857	2.500	2.222	2.00
20		4.61	5.000	4.546	4.166	4.0	3.773	3.572	3.333	2.857	2.500	2.222	2.00
21			4.375	4.545	4.167	4.0	3.774	3.571	3.334	2.857	2.500	2.222	2.00
22				4.546	4.166	4.0	3.773	3.572	3.333	2.857	2.500	2.222	2.00
23				3.977	4.167	4.0	3.774	3.571	3.334	2.857	2.500	2.222	2.00
24					4.166	4.0	3.773	3.572	3.333	2.857	2.500	2.222	2.00
25					3.646	4.0	3.774	3.571	3.334	2.857	2.500	2.222	2.00
26						3.5	3.773	3.572	3.333	2.857	2.500	2.222	2.00
27							3.774	3.571	3.334	2.858	2.500	2.222	2.00
28							1.415	3.572	3.333	2.857	2.500	2.223	2.00
29								3.125	3.334	2.858	2.500	2.222	2.00
30									3.333	2.857	2.500	2.223	2.00
31									2.917	2.858	2.500	2.222	2.00
32										2.857	2.500	2.223	2.00
33										2.858	2.500	2.222	2.00
34										2.857	2.500	2.223	2.00
35										2.858	2.500	2.222	2.00
36										2.500	2.500	2.223	2.00
37											2.500	2.222	2.00
38											2.500	2.223	2.00
39											2.500	2.222	2.00
40											2.500	2.223	2.00
41											2.187	2.222	2.00
42												2.223	2.00
43												2.222	2.00
44												2.223	2.00
45												2.222	2.00
46												1.945	2.00
47–50													2.00
51													1.75

Figure 2.22 continued

Credit: IRS Publication 946, "How to Depreciate Property."

Table A-13. **Straight Line Mid-Month Convention**

Year	Month property placed in service											
	1	2	3	4	5	6	7	8	9	10	11	12
1	2.396%	2.188%	1.979%	1.771%	1.563%	1.354%	1.146%	0.938%	0.729%	0.521%	0.313%	0.104%
2–40	2.500	2.500	2.500	2.500	2.500	2.500	2.500	2.500	2.500	2.500	2.500	2.500
41	0.104	0.312	0.521	0.729	0.937	1.146	1.354	1.562	1.771	1.979	2.187	2.396

Figure 2.23

Credit: IRS Publication 946, "How to Depreciate Property."

Table A-14. **150% Declining Balance Method Half-Year Convention**

Year	Recovery periods in years												
	2.5	3	3.5	4	5	6	6.5	7	7.5	8	8.5	9	9.5
1	30.0%	25.0%	21.43%	18.75%	15.00%	12.50%	11.54%	10.71%	10.00%	9.38%	8.82%	8.33%	7.89%
2	42.0	37.5	33.67	30.47	25.50	21.88	20.41	19.13	18.00	16.99	16.09	15.28	14.54
3	28.0	25.0	22.45	20.31	17.85	16.41	15.70	15.03	14.40	13.81	13.25	12.73	12.25
4		12.5	22.45	20.31	16.66	14.06	13.09	12.25	11.52	11.22	10.91	10.61	10.31
5				10.16	16.66	14.06	13.09	12.25	11.52	10.80	10.19	9.65	9.17
6					8.33	14.06	13.09	12.25	11.52	10.80	10.19	9.64	9.17
7						7.03	13.08	12.25	11.52	10.80	10.18	9.65	9.17
8								6.13	11.52	10.80	10.19	9.64	9.17
9										5.40	10.18	9.65	9.17
10												4.82	9.16

Table A-14. *(Continued)*

Year	Recovery periods in years												
	10	10.5	11	11.5	12	12.5	13	13.5	14	15	16	16.5	17
1	7.50%	7.14%	6.82%	6.52%	6.25%	6.00%	5.77%	5.56%	5.36%	5.00%	4.69%	4.55%	4.41%
2	13.88	13.27	12.71	12.19	11.72	11.28	10.87	10.49	10.14	9.50	8.94	8.68	8.43
3	11.79	11.37	10.97	10.60	10.25	9.93	9.62	9.33	9.05	8.55	8.10	7.89	7.69
4	10.02	9.75	9.48	9.22	8.97	8.73	8.51	8.29	8.08	7.70	7.34	7.17	7.01
5	8.74	8.35	8.18	8.02	7.85	7.69	7.53	7.37	7.22	6.93	6.65	6.52	6.39
6	8.74	8.35	7.98	7.64	7.33	7.05	6.79	6.55	6.44	6.23	6.03	5.93	5.83
7	8.74	8.35	7.97	7.64	7.33	7.05	6.79	6.55	6.32	5.90	5.55	5.39	5.32
8	8.74	8.35	7.98	7.63	7.33	7.05	6.79	6.55	6.32	5.90	5.55	5.39	5.23
9	8.74	8.36	7.97	7.64	7.33	7.04	6.79	6.55	6.32	5.91	5.55	5.39	5.23
10	8.74	8.35	7.98	7.63	7.33	7.05	6.79	6.55	6.32	5.90	5.55	5.39	5.23
11	4.37	8.36	7.97	7.64	7.32	7.04	6.79	6.55	6.32	5.91	5.55	5.39	5.23
12			3.99	7.63	7.33	7.05	6.78	6.55	6.32	5.90	5.55	5.39	5.23
13					3.66	7.04	6.79	6.56	6.32	5.91	5.54	5.38	5.23
14							3.39	6.55	6.31	5.90	5.55	5.39	5.23
15									3.16	5.91	5.54	5.38	5.23
16										2.95	5.55	5.39	5.23
17											2.77	5.38	5.23
18													2.02

Figure 2.24

Credit: IRS Publication 946, "How to Depreciate Property."

Table A-14. *(Continued)*

Year	Recovery periods in years												
	18	19	20	22	24	25	26.5	28	30	35	40	45	50
1	4.17%	3.95%	3.750%	3.409%	3.125%	3.000%	2.830%	2.679%	2.500%	2.143%	1.875%	1.667%	1.500%
2	7.99	7.58	7.219	6.586	6.055	5.820	5.500	5.214	4.875	4.194	3.680	3.278	2.955
3	7.32	6.98	6.677	6.137	5.676	5.471	5.189	4.934	4.631	4.014	3.542	3.169	2.866
4	6.71	6.43	6.177	5.718	5.322	5.143	4.895	4.670	4.400	3.842	3.409	3.063	2.780
5	6.15	5.93	5.713	5.328	4.989	4.834	4.618	4.420	4.180	3.677	3.281	2.961	2.697
6	5.64	5.46	5.285	4.965	4.677	4.544	4.357	4.183	3.971	3.520	3.158	2.862	2.616
7	5.17	5.03	4.888	4.627	4.385	4.271	4.110	3.959	3.772	3.369	3.040	2.767	2.538
8	4.94	4.69	4.522	4.311	4.111	4.015	3.877	3.747	3.584	3.225	2.926	2.674	2.461
9	4.94	4.69	4.462	4.063	3.854	3.774	3.658	3.546	3.404	3.086	2.816	2.585	2.388
10	4.94	4.69	4.461	4.063	3.729	3.584	3.451	3.356	3.234	2.954	2.710	2.499	2.316
11	4.94	4.69	4.462	4.063	3.729	3.583	3.383	3.205	3.072	2.828	2.609	2.416	2.246
12	4.95	4.69	4.461	4.063	3.729	3.584	3.383	3.205	2.994	2.706	2.511	2.335	2.179
13	4.94	4.69	4.462	4.064	3.730	3.583	3.383	3.205	2.994	2.590	2.417	2.257	2.114
14	4.95	4.69	4.461	4.063	3.729	3.584	3.383	3.205	2.994	2.571	2.326	2.182	2.050
15	4.94	4.69	4.462	4.064	3.730	3.583	3.383	3.205	2.994	2.571	2.253	2.110	1.989
16	4.95	4.69	4.461	4.063	3.729	3.584	3.383	3.205	2.994	2.571	2.253	2.039	1.929
17	4.94	4.69	4.462	4.064	3.730	3.583	3.383	3.205	2.994	2.571	2.253	2.005	1.871
18	4.95	4.70	4.461	4.063	3.729	3.584	3.383	3.205	2.994	2.571	2.253	2.005	1.815
19	2.47	4.69	4.462	4.064	3.730	3.583	3.383	3.205	2.994	2.571	2.253	2.005	1.806
20		2.35	4.461	4.063	3.729	3.584	3.384	3.205	2.993	2.571	2.253	2.005	1.806
21			2.231	4.064	3.730	3.583	3.383	3.205	2.994	2.571	2.253	2.005	1.806
22				4.063	3.729	3.584	3.384	3.205	2.993	2.571	2.253	2.005	1.806
23				2.032	3.730	3.583	3.383	3.205	2.994	2.571	2.253	2.005	1.806
24					3.729	3.584	3.384	3.205	2.993	2.571	2.253	2.004	1.806
25					1.865	3.583	3.383	3.205	2.994	2.571	2.253	2.005	1.806
26						1.792	3.384	3.205	2.993	2.571	2.253	2.004	1.806
27							3.383	3.205	2.994	2.571	2.253	2.005	1.806
28								3.205	2.993	2.572	2.253	2.004	1.806
29								1.602	2.994	2.571	2.253	2.005	1.806
30									2.993	2.572	2.253	2.004	1.806
31									1.497	2.571	2.253	2.005	1.806
32										2.572	2.253	2.004	1.806
33										2.571	2.252	2.005	1.806
34										2.572	2.253	2.004	1.806
35										2.571	2.252	2.005	1.806
36										1.286	2.253	2.004	1.806
37											2.252	2.005	1.806
38											2.253	2.004	1.806
39											2.252	2.005	1.806
40											2.253	2.004	1.806
41											1.126	2.005	1.806
42												2.004	1.805
43												2.005	1.806
44												2.004	1.805
45												2.005	1.806
46												1.002	1.805
47													1.806
48													1.805
49													1.806
50													1.805
51													0.903

Figure 2.24 continued

Credit: IRS Publication 946, "How to Depreciate Property."

Table A-15. **150% Declining Balance Method**
Mid-Quarter Convention
Property Placed in Service in First Quarter

Year	Recovery periods in years												
	2.5	**3**	**3.5**	**4**	**5**	**6**	**6.5**	**7**	**7.5**	**8**	**8.5**	**9**	**9.5**
1	52.50%	43.75%	37.50%	32.81%	26.25%	21.88%	20.19%	18.75%	17.50%	16.41%	15.44%	14.58%	13.82%
2	29.23	28.13	26.79	25.20	22.13	19.53	18.42	17.41	16.50	15.67	14.92	14.24	13.61
3	18.27	25.00	21.98	19.76	16.52	14.65	14.17	13.68	13.20	12.74	12.29	11.86	11.46
4		3.12	13.73	19.76	16.52	14.06	13.03	12.16	11.42	10.77	10.20	9.89	9.65
5				2.47	16.52	14.06	13.02	12.16	11.42	10.77	10.19	9.64	9.15
6					2.06	14.06	13.03	12.16	11.41	10.76	10.20	9.65	9.15
7						1.76	8.14	12.16	11.42	10.77	10.19	9.64	9.15
8								1.52	7.13	10.76	10.20	9.65	9.15
9										1.35	6.37	9.64	9.14
10												1.21	5.72

Table A-15. *(Continued)*

Year	Recovery periods in years												
	10	**10.5**	**11**	**11.5**	**12**	**12.5**	**13**	**13.5**	**14**	**15**	**16**	**16.5**	**17**
1	13.13%	12.50%	11.93%	11.41%	10.94%	10.50%	10.10%	9.72%	9.38%	8.75%	8.20%	7.95%	7.72%
2	13.03	12.50	12.01	11.56	11.13	10.74	10.37	10.03	9.71	9.13	8.61	8.37	8.14
3	11.08	10.71	10.37	10.05	9.74	9.45	9.18	8.92	8.67	8.21	7.80	7.61	7.42
4	9.41	9.18	8.96	8.74	8.52	8.32	8.12	7.93	7.74	7.39	7.07	6.92	6.77
5	8.71	8.32	7.96	7.64	7.46	7.32	7.18	7.04	6.91	6.65	6.41	6.29	6.17
6	8.71	8.32	7.96	7.64	7.33	7.04	6.78	6.53	6.31	5.99	5.80	5.71	5.63
7	8.71	8.32	7.96	7.64	7.33	7.04	6.77	6.54	6.31	5.90	5.54	5.38	5.23
8	8.71	8.32	7.96	7.64	7.33	7.04	6.78	6.53	6.31	5.91	5.54	5.38	5.23
9	8.71	8.32	7.96	7.64	7.33	7.04	6.77	6.54	6.31	5.90	5.54	5.38	5.23
10	8.71	8.31	7.97	7.63	7.32	7.04	6.78	6.53	6.31	5.91	5.54	5.38	5.23
11	1.09	5.20	7.96	7.64	7.33	7.04	6.77	6.54	6.31	5.90	5.54	5.38	5.23
12			1.00	4.77	7.32	7.03	6.78	6.53	6.31	5.91	5.54	5.38	5.22
13					0.92	4.40	6.77	6.54	6.32	5.90	5.54	5.38	5.23
14							0.85	4.08	6.31	5.91	5.55	5.38	5.22
15									0.79	5.90	5.54	5.38	5.23
16										0.74	5.55	5.37	5.22
17											0.69	3.36	5.23
18													0.65

Figure 2.25
Credit: IRS Publication 946, "How to Depreciate Property."

Table A-15. *(Continued)*

Year	Recovery periods in years												
	18	19	20	22	24	25	26.5	28	30	35	40	45	50
1	7.29%	6.91%	6.563%	5.966%	5.469%	5.250%	4.953%	4.688%	4.375%	3.750%	3.281%	2.917%	2.625%
2	7.73	7.35	7.008	6.411	5.908	5.685	5.380	5.106	4.781	4.125	3.627	3.236	2.921
3	7.08	6.77	6.482	5.974	5.539	5.344	5.075	4.832	4.542	3.948	3.491	3.128	2.834
4	6.49	6.23	5.996	5.567	5.193	5.023	4.788	4.574	4.315	3.779	3.360	3.024	2.749
5	5.95	5.74	5.546	5.187	4.868	4.722	4.517	4.329	4.099	3.617	3.234	2.923	2.666
6	5.45	5.29	5.130	4.834	4.564	4.439	4.262	4.097	3.894	3.462	3.113	2.826	2.586
7	5.00	4.87	4.746	4.504	4.279	4.172	4.020	3.877	3.700	3.314	2.996	2.732	2.509
8	4.94	4.69	4.459	4.197	4.011	3.922	3.793	3.669	3.515	3.172	2.884	2.640	2.433
9	4.95	4.69	4.459	4.061	3.761	3.687	3.578	3.473	3.339	3.036	2.776	2.552	2.360
10	4.94	4.69	4.459	4.061	3.729	3.582	3.383	3.287	3.172	2.906	2.671	2.467	2.290
11	4.95	4.69	4.459	4.061	3.729	3.582	3.384	3.204	3.013	2.781	2.571	2.385	2.221
12	4.94	4.69	4.460	4.061	3.730	3.582	3.383	3.204	2.994	2.662	2.475	2.306	2.154
13	4.95	4.69	4.459	4.061	3.729	3.582	3.384	3.204	2.994	2.571	2.382	2.229	2.090
14	4.94	4.69	4.460	4.061	3.730	3.582	3.383	3.204	2.994	2.571	2.293	2.154	2.027
15	4.95	4.68	4.459	4.061	3.729	3.582	3.384	3.204	2.994	2.571	2.252	2.083	1.966
16	4.94	4.69	4.460	4.061	3.730	3.582	3.383	3.204	2.994	2.571	2.252	2.013	1.907
17	4.95	4.68	4.459	4.061	3.729	3.582	3.384	3.204	2.994	2.571	2.253	2.005	1.850
18	4.94	4.69	4.460	4.061	3.730	3.582	3.383	3.204	2.994	2.571	2.252	2.005	1.806
19	0.62	4.68	4.459	4.061	3.729	3.581	3.384	3.204	2.994	2.571	2.253	2.005	1.806
20		0.59	4.460	4.060	3.730	3.582	3.383	3.204	2.994	2.571	2.252	2.005	1.806
21			0.557	4.061	3.729	3.581	3.384	3.203	2.993	2.571	2.253	2.005	1.806
22				4.060	3.730	3.582	3.383	3.204	2.994	2.571	2.252	2.005	1.806
23				0.508	3.729	3.581	3.384	3.203	2.993	2.571	2.253	2.005	1.806
24					3.730	3.582	3.383	3.204	2.994	2.570	2.252	2.005	1.806
25					0.466	3.581	3.384	3.203	2.993	2.571	2.253	2.004	1.806
26						0.448	3.383	3.204	2.994	2.570	2.252	2.005	1.806
27							2.115	3.203	2.993	2.571	2.253	2.004	1.806
28								3.204	2.994	2.570	2.252	2.005	1.805
29								0.400	2.993	2.571	2.253	2.004	1.806
30									2.994	2.570	2.252	2.005	1.805
31									0.374	2.571	2.253	2.004	1.806
32										2.570	2.252	2.005	1.805
33										2.571	2.253	2.004	1.806
34										2.570	2.252	2.005	1.805
35										2.571	2.253	2.004	1.806
36										0.321	2.252	2.005	1.805
37											2.253	2.004	1.806
38											2.252	2.005	1.805
39											2.253	2.004	1.806
40											2.252	2.005	1.805
41											0.282	2.004	1.806
42												2.005	1.805
43												2.004	1.806
44												2.005	1.805
45												2.004	1.806
46												0.251	1.805
47													1.806
48													1.805
49													1.806
50													1.805
51													0.226

Figure 2.25 continued

Credit: IRS Publication 946, "How to Depreciate Property."

Table A-16. **150% Declining Balance Method**
Mid-Quarter Convention
Property Placed in Service in Second Quarter

Year	Recovery periods in years												
	2.5	**3**	**3.5**	**4**	**5**	**6**	**6.5**	**7**	**7.5**	**8**	**8.5**	**9**	**9.5**
1	37.50%	31.25%	26.79%	23.44%	18.75%	15.63%	14.42%	13.39%	12.50%	11.72%	11.03%	10.42%	9.87%
2	37.50	34.38	31.38	28.71	24.38	21.09	19.75	18.56	17.50	16.55	15.70	14.93	14.23
3	25.00	25.00	22.31	20.15	17.06	15.82	15.19	14.58	14.00	13.45	12.93	12.44	11.98
4		9.37	19.52	20.15	16.76	14.06	13.07	12.22	11.49	10.93	10.65	10.37	10.09
5				7.55	16.76	14.06	13.07	12.22	11.49	10.82	10.19	9.64	9.16
6					6.29	14.07	13.07	12.22	11.49	10.82	10.19	9.65	9.16
7						5.27	11.43	12.23	11.48	10.83	10.19	9.64	9.16
8								4.58	10.05	10.82	10.20	9.65	9.17
9										4.06	8.92	9.64	9.16
10												3.62	8.02

Table A-16. *(Continued)*

Year	Recovery periods in years												
	10	**10.5**	**11**	**11.5**	**12**	**12.5**	**13**	**13.5**	**14**	**15**	**16**	**16.5**	**17**
1	9.38%	8.93%	8.52%	8.15%	7.81%	7.50%	7.21%	6.94%	6.70%	6.25%	5.86%	5.68%	5.51%
2	13.59	13.01	12.47	11.98	11.52	11.10	10.71	10.34	10.00	9.38	8.83	8.57	8.34
3	11.55	11.15	10.77	10.42	10.08	9.77	9.47	9.19	8.92	8.44	8.00	7.80	7.60
4	9.82	9.56	9.31	9.06	8.82	8.60	8.38	8.17	7.97	7.59	7.25	7.09	6.93
5	8.73	8.34	8.04	7.88	7.72	7.56	7.41	7.26	7.12	6.83	6.57	6.44	6.32
6	8.73	8.34	7.98	7.64	7.33	7.04	6.78	6.55	6.35	6.15	5.95	5.86	5.76
7	8.73	8.34	7.98	7.64	7.33	7.04	6.79	6.55	6.32	5.91	5.55	5.38	5.25
8	8.73	8.34	7.98	7.64	7.33	7.05	6.78	6.55	6.32	5.90	5.55	5.39	5.23
9	8.73	8.34	7.99	7.64	7.33	7.04	6.79	6.54	6.32	5.91	5.55	5.38	5.23
10	8.73	8.35	7.98	7.63	7.33	7.05	6.78	6.55	6.32	5.90	5.54	5.39	5.23
11	3.28	7.30	7.99	7.64	7.33	7.04	6.79	6.54	6.32	5.91	5.55	5.38	5.23
12			2.99	6.68	7.32	7.05	6.78	6.55	6.32	5.90	5.54	5.39	5.23
13					2.75	6.16	6.79	6.54	6.32	5.91	5.55	5.38	5.24
14							2.54	5.73	6.33	5.90	5.54	5.39	5.23
15									2.37	5.91	5.55	5.38	5.24
16										2.21	5.54	5.39	5.23
17											2.08	4.71	5.24
18													1.96

Figure 2.26
Credit: IRS Publication 946, "How to Depreciate Property."

Table A-16. *(Continued)*

Year	Recovery periods in years												
	18	19	20	22	24	25	26.5	28	30	35	40	45	50
1	5.21%	4.93%	4.688%	4.261%	3.906%	3.750%	3.538%	3.348%	3.125%	2.679%	2.344%	2.083%	1.875%
2	7.90	7.51	7.148	6.528	6.006	5.775	5.460	5.178	4.844	4.171	3.662	3.264	2.944
3	7.24	6.91	6.612	6.083	5.631	5.429	5.151	4.900	4.602	3.992	3.525	3.155	2.855
4	6.64	6.37	6.116	5.668	5.279	5.103	4.859	4.638	4.371	3.821	3.393	3.050	2.770
5	6.08	5.86	5.658	5.281	4.949	4.797	4.584	4.389	4.153	3.657	3.265	2.948	2.687
6	5.58	5.40	5.233	4.921	4.639	4.509	4.325	4.154	3.945	3.501	3.143	2.850	2.606
7	5.11	4.98	4.841	4.586	4.349	4.238	4.080	3.932	3.748	3.351	3.025	2.755	2.528
8	4.94	4.69	4.478	4.273	4.078	3.984	3.849	3.721	3.561	3.207	2.912	2.663	2.452
9	4.94	4.69	4.463	4.063	3.823	3.745	3.631	3.522	3.383	3.069	2.802	2.574	2.378
10	4.95	4.69	4.463	4.063	3.729	3.583	3.426	3.333	3.213	2.938	2.697	2.489	2.307
11	4.94	4.69	4.463	4.062	3.729	3.583	3.384	3.205	3.053	2.812	2.596	2.406	2.238
12	4.95	4.69	4.463	4.063	3.729	3.583	3.383	3.205	2.994	2.692	2.499	2.325	2.171
13	4.94	4.69	4.463	4.062	3.730	3.583	3.384	3.205	2.994	2.576	2.405	2.248	2.106
14	4.95	4.69	4.463	4.063	3.729	3.583	3.383	3.205	2.994	2.571	2.315	2.173	2.042
15	4.94	4.69	4.462	4.062	3.730	3.583	3.384	3.205	2.994	2.571	2.253	2.101	1.981
16	4.95	4.69	4.463	4.063	3.729	3.583	3.383	3.204	2.994	2.571	2.253	2.031	1.922
17	4.94	4.69	4.462	4.062	3.730	3.583	3.384	3.205	2.994	2.571	2.253	2.005	1.864
18	4.95	4.69	4.463	4.063	3.729	3.583	3.383	3.204	2.993	2.571	2.253	2.005	1.808
19	1.85	4.69	4.462	4.062	3.730	3.583	3.384	3.205	2.994	2.571	2.253	2.005	1.806
20		1.76	4.463	4.063	3.729	3.583	3.383	3.204	2.993	2.571	2.253	2.005	1.806
21			1.673	4.062	3.730	3.583	3.384	3.205	2.994	2.572	2.253	2.005	1.806
22				4.063	3.729	3.583	3.383	3.204	2.993	2.571	2.253	2.005	1.806
23				1.523	3.730	3.583	3.384	3.205	2.994	2.572	2.253	2.004	1.806
24					3.729	3.582	3.383	3.204	2.993	2.571	2.253	2.005	1.806
25					1.399	3.583	3.384	3.205	2.994	2.572	2.253	2.004	1.806
26						1.343	3.383	3.204	2.993	2.571	2.253	2.005	1.806
27							2.961	3.205	2.994	2.572	2.253	2.004	1.806
28								3.204	2.993	2.571	2.253	2.005	1.806
29								1.202	2.994	2.572	2.253	2.004	1.806
30									2.993	2.571	2.252	2.005	1.806
31									1.123	2.572	2.253	2.004	1.806
32										2.571	2.252	2.005	1.806
33										2.572	2.253	2.004	1.806
34										2.571	2.252	2.005	1.806
35										2.572	2.253	2.004	1.806
36										0.964	2.252	2.005	1.806
37											2.253	2.004	1.806
38											2.252	2.005	1.806
39											2.253	2.004	1.806
40											2.252	2.005	1.806
41											0.845	2.004	1.806
42												2.005	1.806
43												2.004	1.806
44												2.005	1.806
45												2.004	1.805
46												0.752	1.806
47													1.805
48													1.806
49													1.805
50													1.806
51													0.677

Figure 2.26 continued

Credit: IRS Publication 946, "How to Depreciate Property."

Table A-17. **150% Declining Balance Method**
Mid-Quarter Convention
Property Placed in Service in Third Quarter

Year	Recovery periods in years												
	2.5	**3**	**3.5**	**4**	**5**	**6**	**6.5**	**7**	**7.5**	**8**	**8.5**	**9**	**9.5**
1	22.50%	18.75%	16.07%	14.06%	11.25%	9.38%	8.65%	8.04%	7.50%	7.03%	6.62%	6.25%	5.92%
2	46.50	40.63	35.97	32.23	26.63	22.66	21.08	19.71	18.50	17.43	16.48	15.63	14.85
3	27.56	25.00	22.57	20.46	18.64	16.99	16.22	15.48	14.80	14.16	13.57	13.02	12.51
4	3.44	15.62	22.57	20.46	16.56	14.06	13.10	12.27	11.84	11.51	11.18	10.85	10.53
5			2.82	12.79	16.57	14.06	13.10	12.28	11.48	10.78	10.18	9.64	9.17
6					10.35	14.06	13.11	12.27	11.48	10.78	10.17	9.65	9.17
7						8.79	13.10	12.28	11.48	10.78	10.18	9.64	9.18
8							1.64	7.67	11.48	10.79	10.17	9.65	9.17
9									1.44	6.74	10.18	9.64	9.18
10											1.27	6.03	9.17
11													1.15

Table A-17. *(Continued)*

Year	Recovery periods in years												
	10	**10.5**	**11**	**11.5**	**12**	**12.5**	**13**	**13.5**	**14**	**15**	**16**	**16.5**	**17**
1	5.63%	5.36%	5.11%	4.89%	4.69%	4.50%	4.33%	4.17%	4.02%	3.75%	3.52%	3.41%	3.31%
2	14.16	13.52	12.94	12.41	11.91	11.46	11.04	10.65	10.28	9.63	9.05	8.78	8.53
3	12.03	11.59	11.18	10.79	10.43	10.08	9.77	9.46	9.18	8.66	8.20	7.98	7.78
4	10.23	9.93	9.65	9.38	9.12	8.88	8.64	8.41	8.20	7.80	7.43	7.26	7.09
5	8.75	8.51	8.33	8.16	7.98	7.81	7.64	7.48	7.32	7.02	6.73	6.60	6.47
6	8.75	8.34	7.97	7.63	7.33	7.05	6.79	6.65	6.54	6.31	6.10	6.00	5.90
7	8.75	8.34	7.97	7.63	7.33	7.05	6.79	6.55	6.31	5.90	5.55	5.45	5.38
8	8.74	8.34	7.97	7.63	7.33	7.05	6.79	6.54	6.31	5.90	5.55	5.38	5.23
9	8.75	8.34	7.97	7.63	7.33	7.05	6.79	6.55	6.32	5.91	5.55	5.39	5.23
10	8.74	8.34	7.97	7.63	7.32	7.05	6.79	6.54	6.31	5.90	5.55	5.38	5.23
11	5.47	8.35	7.96	7.63	7.33	7.05	6.79	6.55	6.32	5.91	5.55	5.39	5.23
12		1.04	4.98	7.64	7.32	7.04	6.80	6.54	6.31	5.90	5.55	5.38	5.23
13				0.95	4.58	7.05	6.79	6.55	6.32	5.91	5.55	5.39	5.22
14						0.88	4.25	6.54	6.31	5.90	5.55	5.38	5.23
15								0.82	3.95	5.91	5.55	5.39	5.22
16										3.69	5.55	5.38	5.23
17											3.47	5.39	5.22
18												0.67	3.27

Figure 2.27
Credit: IRS Publication 946, "How to Depreciate Property."

Table A-17. *(Continued)*

Year	Recovery periods in years												
	18	19	20	22	24	25	26.5	28	30	35	40	45	50
1	3.13%	2.96%	2.813%	2.557%	2.344%	2.250%	2.123%	2.009%	1.875%	1.607%	1.406%	1.250%	1.125%
2	8.07	7.66	7.289	6.644	6.104	5.865	5.540	5.250	4.906	4.217	3.697	3.292	2.966
3	7.40	7.06	6.742	6.191	5.722	5.513	5.227	4.968	4.661	4.036	3.559	3.182	2.877
4	6.78	6.50	6.237	5.769	5.364	5.182	4.931	4.702	4.428	3.863	3.425	3.076	2.791
5	6.22	5.99	5.769	5.375	5.029	4.871	4.652	4.450	4.207	3.698	3.297	2.973	2.707
6	5.70	5.51	5.336	5.009	4.715	4.579	4.388	4.212	3.996	3.539	3.173	2.874	2.626
7	5.23	5.08	4.936	4.667	4.420	4.304	4.140	3.986	3.796	3.387	3.054	2.778	2.547
8	4.94	4.69	4.566	4.349	4.144	4.046	3.906	3.773	3.607	3.242	2.940	2.686	2.471
9	4.94	4.69	4.460	4.064	3.885	3.803	3.685	3.571	3.426	3.103	2.829	2.596	2.397
10	4.94	4.69	4.460	4.064	3.729	3.584	3.476	3.379	3.255	2.970	2.723	2.510	2.325
11	4.94	4.69	4.460	4.064	3.730	3.584	3.383	3.205	3.092	2.843	2.621	2.426	2.255
12	4.95	4.69	4.460	4.064	3.729	3.584	3.383	3.205	2.994	2.721	2.523	2.345	2.187
13	4.94	4.69	4.461	4.064	3.730	3.584	3.383	3.205	2.994	2.605	2.428	2.267	2.122
14	4.95	4.69	4.460	4.064	3.729	3.584	3.383	3.205	2.994	2.571	2.337	2.192	2.058
15	4.94	4.70	4.461	4.064	3.730	3.584	3.383	3.205	2.994	2.571	2.253	2.118	1.996
16	4.95	4.69	4.460	4.064	3.729	3.584	3.383	3.206	2.994	2.571	2.253	2.048	1.937
17	4.94	4.70	4.461	4.064	3.730	3.584	3.383	3.205	2.994	2.571	2.253	2.005	1.878
18	4.95	4.69	4.460	4.065	3.729	3.584	3.383	3.206	2.994	2.571	2.253	2.005	1.822
19	3.09	4.70	4.461	4.064	3.730	3.584	3.383	3.205	2.994	2.571	2.253	2.005	1.806
20		2.93	4.460	4.065	3.729	3.584	3.383	3.206	2.993	2.571	2.253	2.005	1.806
21			2.788	4.064	3.730	3.585	3.383	3.205	2.994	2.571	2.253	2.005	1.806
22				4.065	3.729	3.584	3.383	3.206	2.993	2.571	2.253	2.005	1.806
23				2.540	3.730	3.585	3.383	3.205	2.994	2.571	2.253	2.005	1.806
24					3.729	3.584	3.383	3.206	2.993	2.571	2.253	2.005	1.806
25					2.331	3.585	3.382	3.205	2.994	2.571	2.253	2.004	1.806
26						2.240	3.383	3.206	2.993	2.571	2.253	2.005	1.806
27							3.382	3.205	2.994	2.571	2.253	2.004	1.806
28							0.423	3.206	2.993	2.571	2.253	2.005	1.806
29								2.003	2.994	2.571	2.253	2.004	1.806
30									2.993	2.571	2.253	2.005	1.806
31									1.871	2.571	2.253	2.004	1.806
32										2.571	2.253	2.005	1.806
33										2.571	2.253	2.004	1.806
34										2.571	2.253	2.005	1.806
35										2.571	2.253	2.004	1.806
36										1.607	2.253	2.005	1.806
37											2.253	2.004	1.805
38											2.254	2.005	1.806
39											2.253	2.004	1.805
40											2.254	2.005	1.806
41											1.408	2.004	1.805
42												2.005	1.806
43												2.004	1.805
44												2.005	1.806
45												2.004	1.805
46												1.253	1.806
47													1.805
48													1.806
49													1.805
50													1.806
51													1.128

Figure 2.27 continued

Credit: IRS Publication 946, "How to Depreciate Property."

Table A-18. **150% Declining Balance Method**
Mid-Quarter Convention
Property Placed in Service in Fourth Quarter

Year	Recovery periods in years												
	2.5	**3**	**3.5**	**4**	**5**	**6**	**6.5**	**7**	**7.5**	**8**	**8.5**	**9**	**9.5**
1	7.50%	6.25%	5.36%	4.69%	3.75%	3.13%	2.88%	2.68%	2.50%	2.34%	2.21%	2.08%	1.97%
2	55.50	46.88	40.56	35.74	28.88	24.22	22.41	20.85	19.50	18.31	17.26	16.32	15.48
3	26.91	25.00	23.18	22.34	20.21	18.16	17.24	16.39	15.60	14.88	14.21	13.60	13.03
4	10.09	21.87	22.47	19.86	16.40	14.06	13.26	12.87	12.48	12.09	11.70	11.33	10.98
5			8.43	17.37	16.41	14.06	13.10	12.18	11.41	10.74	10.16	9.65	9.24
6					14.35	14.06	13.10	12.18	11.41	10.75	10.16	9.65	9.17
7						12.31	13.10	12.19	11.41	10.74	10.16	9.64	9.17
8							4.91	10.66	11.41	10.75	10.16	9.65	9.17
9									4.28	9.40	10.17	9.64	9.17
10											3.81	8.44	9.18
11													3.44

Table A-18. *(Continued)*

Year	Recovery periods in years												
	10	**10.5**	**11**	**11.5**	**12**	**12.5**	**13**	**13.5**	**14**	**15**	**16**	**16.5**	**17**
1	1.88%	1.79%	1.70%	1.63%	1.56%	1.50%	1.44%	1.39%	1.34%	1.25%	1.17%	1.14%	1.10%
2	14.72	14.03	13.40	12.83	12.31	11.82	11.37	10.96	10.57	9.88	9.27	8.99	8.73
3	12.51	12.03	11.58	11.16	10.77	10.40	10.06	9.74	9.44	8.89	8.40	8.17	7.96
4	10.63	10.31	10.00	9.70	9.42	9.15	8.90	8.66	8.43	8.00	7.61	7.43	7.25
5	9.04	8.83	8.63	8.44	8.24	8.06	7.87	7.69	7.52	7.20	6.90	6.75	6.61
6	8.72	8.32	7.95	7.63	7.33	7.09	6.96	6.84	6.72	6.48	6.25	6.14	6.03
7	8.72	8.31	7.96	7.63	7.33	7.05	6.78	6.53	6.31	5.90	5.66	5.58	5.50
8	8.72	8.32	7.95	7.62	7.33	7.05	6.78	6.53	6.31	5.90	5.54	5.38	5.22
9	8.72	8.31	7.96	7.63	7.33	7.05	6.78	6.53	6.31	5.90	5.54	5.38	5.23
10	8.71	8.32	7.95	7.62	7.32	7.05	6.78	6.54	6.31	5.91	5.54	5.38	5.22
11	7.63	8.31	7.96	7.63	7.33	7.05	6.78	6.53	6.31	5.90	5.54	5.38	5.23
12		3.12	6.96	7.62	7.32	7.04	6.78	6.54	6.30	5.91	5.55	5.38	5.22
13				2.86	6.41	7.05	6.78	6.53	6.31	5.90	5.54	5.38	5.23
14						2.64	5.94	6.54	6.30	5.91	5.55	5.38	5.22
15								2.45	5.52	5.90	5.54	5.37	5.23
16										5.17	5.55	5.38	5.22
17											4.85	5.37	5.23
18												2.02	4.57

Figure 2.28
Credit: IRS Publication 946, "How to Depreciate Property."

Table A-18. *(Continued)*

Year	Recovery periods in years												
	18	19	20	22	24	25	26.5	28	30	35	40	45	50
1	1.04%	0.99%	0.938%	0.852%	0.781%	0.750%	0.708%	0.670%	0.625%	0.536%	0.469%	0.417%	0.375%
2	8.25	7.82	7.430	6.760	6.201	5.955	5.620	5.321	4.969	4.263	3.732	3.319	2.989
3	7.56	7.20	6.872	6.299	5.814	5.598	5.302	5.036	4.720	4.080	3.592	3.209	2.899
4	6.93	6.63	6.357	5.870	5.450	5.262	5.002	4.766	4.484	3.905	3.458	3.102	2.812
5	6.35	6.11	5.880	5.469	5.110	4.946	4.719	4.511	4.260	3.738	3.328	2.998	2.728
6	5.82	5.63	5.439	5.097	4.790	4.649	4.452	4.269	4.047	3.578	3.203	2.898	2.646
7	5.34	5.18	5.031	4.749	4.491	4.370	4.200	4.041	3.845	3.424	3.083	2.802	2.567
8	4.94	4.77	4.654	4.425	4.210	4.108	3.962	3.824	3.653	3.278	2.968	2.708	2.490
9	4.94	4.69	4.458	4.124	3.947	3.862	3.738	3.619	3.470	3.137	2.856	2.618	2.415
10	4.94	4.69	4.458	4.062	3.730	3.630	3.526	3.426	3.296	3.003	2.749	2.531	2.342
11	4.95	4.69	4.458	4.062	3.729	3.582	3.383	3.242	3.132	2.874	2.646	2.447	2.272
12	4.94	4.69	4.458	4.062	3.730	3.582	3.382	3.204	2.994	2.751	2.547	2.365	2.204
13	4.95	4.69	4.458	4.062	3.729	3.582	3.383	3.204	2.994	2.633	2.451	2.286	2.138
14	4.94	4.69	4.458	4.061	3.730	3.582	3.382	3.204	2.994	2.570	2.359	2.210	2.074
15	4.95	4.69	4.458	4.062	3.729	3.582	3.383	3.204	2.994	2.571	2.271	2.136	2.011
16	4.94	4.69	4.458	4.061	3.730	3.583	3.382	3.204	2.994	2.570	2.253	2.065	1.951
17	4.95	4.68	4.458	4.062	3.729	3.582	3.383	3.204	2.994	2.571	2.253	2.005	1.893
18	4.94	4.69	4.459	4.061	3.730	3.583	3.382	3.204	2.994	2.570	2.253	2.005	1.836
19	4.33	4.68	4.458	4.062	3.729	3.582	3.383	3.204	2.993	2.571	2.253	2.005	1.806
20		4.10	4.459	4.061	3.730	3.583	3.382	3.204	2.994	2.570	2.253	2.005	1.806
21			3.901	4.062	3.729	3.582	3.383	3.204	2.993	2.571	2.253	2.005	1.806
22				4.061	3.730	3.583	3.382	3.204	2.994	2.570	2.253	2.005	1.806
23				3.554	3.729	3.582	3.383	3.205	2.993	2.571	2.253	2.005	1.806
24					3.730	3.583	3.382	3.204	2.994	2.570	2.253	2.005	1.805
25					3.263	3.582	3.383	3.205	2.993	2.571	2.253	2.005	1.806
26						3.135	3.382	3.204	2.994	2.570	2.252	2.005	1.805
27							3.383	3.205	2.993	2.571	2.253	2.004	1.806
28							1.268	3.204	2.994	2.570	2.252	2.005	1.805
29								2.804	2.993	2.571	2.253	2.004	1.806
30									2.994	2.570	2.252	2.005	1.805
31									2.619	2.571	2.253	2.004	1.806
32										2.570	2.252	2.005	1.805
33										2.571	2.253	2.004	1.806
34										2.570	2.252	2.005	1.805
35										2.571	2.253	2.004	1.806
36										2.249	2.252	2.005	1.805
37											2.253	2.004	1.806
38											2.252	2.005	1.805
39											2.253	2.004	1.806
40											2.252	2.005	1.805
41											1.971	2.004	1.806
42												2.005	1.805
43												2.004	1.806
44												2.005	1.805
45												2.004	1.806
46												1.754	1.805
47													1.806
48													1.805
49													1.806
50													1.805
51													1.580

Figure 2.28 continued

Credit: IRS Publication 946, "How to Depreciate Property."

Chapter 3 The Legal Framework for Cost Segregation

All cost segregation studies should have a solid foundation on tax precedents that interpret both tax laws (Internal Revenue Code) and income tax regulations. These precedents, together with the particulars of a client's business operations, will determine the amount and type of cost data that must be compiled or generated by a qualified estimator.

The amount of tax deferral that can be claimed by the taxpayer is supported by detailed cost analysis that accompanies the claim, and the laws and interpretations behind it. In the event that the Internal Revenue Service examines a tax return that includes a schedule for depreciation of fixed assets, detailed cost segregation studies can help substantiate asset classifications and the costs of tangible personal property claimed.

Historically, cost segregation has been addressed at many levels, including:

- Internal Revenue Code
- U.S. Treasury Department regulations
- Tax Court and other court cases
- Revenue procedures
- Various Letter Rulings (internal IRS memos and private Letter Rulings)

This chapter will review the ways in which each of these sources controls and affects the application of cost segregation.

Internal Revenue Code

The U.S. tax system is one of voluntary compliance. Taxpayers and professionals interpret the Internal Revenue Code (IRC) as needed to determine the correct amount of tax due. The IRC is found in Title 26 of the United States Code collection of laws. Subtitle A of Title 26 governs income tax.

The section dealing with the IRC was developed by the U.S. Congress and is updated on an ongoing basis. When a new tax law is passed, committees in both houses compile reports and provide public guidance. Once signed by the President, the bill becomes law and is available for implementation and interpretation.

The section of the Internal Revenue Code that is the starting point for cost segregation is 167(a). This section outlines deductions for depreciation and is designed to address questions related to cost segregation studies: What is depreciation? How is it calculated? Can any piece of property be depreciated? When can it be deducted, and by whom?

IRC Sections 162 and 212 allow for income deductions for ordinary and necessary business expenses, but not for expenditures for assets.

Buildings, land improvements, and tangible personal property items are often usable in the taxpayer's trade or business activities beyond the current year. Over time, however, these assets are subject to wear and tear, and, in some cases, can become obsolete because of changes in an industry or technology. The IRC provides for this loss in value by way of a deduction, known as *depreciation,* covered in Sections 167 and 168 (calculation methods). The current system of depreciation used by most taxpayers is referred to as the Modified Accelerated Cost Recovery System (MACRS), although other methods of depreciation may be used, including:

- The straight-line method
- The 150% declining balance method
- Other methods not expressed in terms of years
  - –Units of production method
  - –Income forecast method

Taxpayers may use any of these methods, as long as they result in a clear reflection of income. *(See Chapter 2 for further explanations of these methods.)*

MACRS

MACRS depreciation has two subsets of calculation methods, the General Depreciation System (GDS) and the Alternative Depreciation System (ADS). Classifying assets under these systems relies heavily on the distinction between real and tangible personal property under the IRC. To make this distinction, it is necessary to analyze Sections 1245, 1250, 38, and 48. Sections 1245 and 1250 provide definitions of tangible personal property, buildings, and "inherently permanent improvements to land." Sections 38 and 48 explain the Investment Tax Credit (allowed from 1962 until its repeal in 1986) and define qualifying property. Property that qualified (this credit is no longer allowed) for the credit has become synonymous with the term *tangible personal property. (See Chapter 2, "Income Taxes & Depreciation Deductions," for more on GDS and ADS.)*

Many other sections of the IRC cover the basis of property, types of property acquisitions, and other matters related to cost segregation. *(These sections will be discussed in Chapter 4, "Tax Cases, IRS Revenue Rulings, & Revenue Procedures.")*

U.S. Treasury Department Regulations

The U.S. Treasury Department regulations attempt to further explain how the laws in the Internal Revenue Code apply to individual situations. *Legislative regulations* are required or requested at the direction of Congress. The Commissioner of the Internal Revenue Service may also issue regulations, commonly referred to as *interpretive regulations.* U.S. Treasury Department regulations provide examples of how the IRS interprets the application of the law, and how taxpayers are expected to apply it to individual circumstances.

When cited, regulations usually appear with the number "1" followed by a decimal and the related code section number, according to topics that are covered. For example, regulations pertaining to the allowance of a deduction for depreciation in Section 167 are Treasury Regulations 1.167-1 or 1.167-2. Property that qualifies for the Investment Tax Credit is discussed in Treasury Regulations 1.48-1(a) and sections that follow.

Disputes

Taxpayers and the IRS may not always agree on the application of the Internal Revenue Code sections and the related regulations. The differences, if they remain unresolved, can be taken to the Tax Court (the highest level under the IRS) for problem resolution. If either party is dissatisfied with the outcome, the case can be heard by the federal court system, although cost constraints sometimes prevent the average taxpayer from taking that step. In the case of the *Hospital Corporation of America*, in which the taxpayer disagreed with the IRS on applicable rules for the purpose of classifying property, the tax dollars at stake were close to $700,000,000. A dispute concerning a small tax liability may not be worth the costs of litigating the case.

The U.S. Tax Court, U.S. District Courts, U.S. Circuit Courts of Appeals, & the U.S. Supreme Court

The cost segregation practitioner should have some knowledge of the court system and how it functions. Being aware of precedents set by prior cases—and how they apply and support a client's case—is essential. The value and applicability of prior cases will depend on the type of case, where it was decided, and the decision that was issued. Agreement to a similar case by the Internal Revenue Service is, of course, helpful to your cause, whereas another case that continues to be disputed may not be helpful.

The higher a case goes in the U.S. judiciary system, the greater the potential that it will establish a precedent. Most tax-related cases begin at the Tax Court level, and appeals are generally made to the Circuit Court of Appeals for the circuit in which the taxpayer's main activities are conducted. For example, a business located in Rhode Island would appeal a

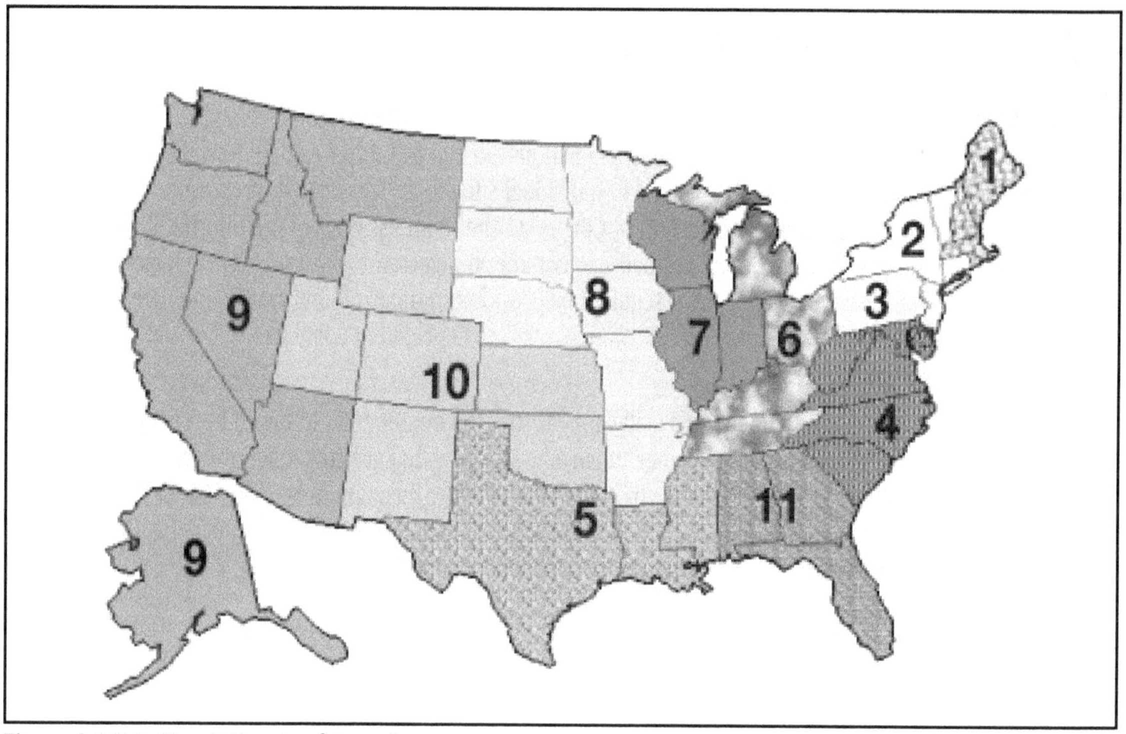

Figure 3.1 U.S. Circuit Courts of Appeals

decision of the Tax Court to the First Circuit Court of Appeals (usually designated in research material as CA-1). Figure 3.1 shows the jurisdictions of each of the 11 U.S. Circuit Courts of Appeals.

Sometimes Circuit Courts arrive at different decisions. Cost segregation practitioners should give more weight to cases decided in their own Circuit Courts than to cases from others, especially when there are discrepancies among them. The Tax Court will follow the case history in the circuit of the taxpayer. When there are discrepancies in similarly situated taxpayers' cases, the U.S. Supreme Court may be asked to hear the cases, a process known as "granting *Certiorari.*" Since the Supreme Court is the highest court, a case heard there should be given the greatest amount of weight over any other cases.

The U.S. Tax Court

The U.S. Tax Court is a federal court of record established by the U.S. Congress under Article I of the U.S. Constitution. Congress created the Tax Court to provide a judicial forum in which affected persons could dispute tax deficiencies determined by the Commissioner of the Internal Revenue Service prior to paying the disputed amounts. The jurisdiction of the Tax Court includes the authority to hear tax disputes concerning:

- Notices of deficiency
- Notices of transferee liability
- Certain types of declaratory judgment
- Adjustment and readjustment of partnership items
- Review of the failure to abate interest
- Administrative costs
- Worker classification

- Relief from joint and several liability on a joint return
- Review of certain collection actions

The Tax Court is composed of 19 members appointed by the President. Although the court is located in Washington, D.C., judges travel nationwide to conduct trials in various cities.

U.S. District Courts

There are 89 districts in the 50 states (listed in Title 28 of the U.S. Code, Sections 81-144) in addition to the five districts in Puerto Rico, the Virgin Islands, the District of Columbia, Guam, and the Northern Mariana Islands. Some states, like Alaska, have a single judicial district, while others, like California, have multiple judicial districts. The number of judgeships allotted to each district is set forth in Title 28 of the U.S. Code, Section 133.

The U.S. Circuit Courts of Appeals

There are 13 judicial circuits, each with a Court of Appeals. The smallest court is the First Circuit with six judgeships, and the largest court is the Ninth Circuit, with 28 judgeships. A list of the states that compose each circuit and the number of judgeships in each circuit are set forth in Title 28 of the U.S. Code, Section 44.

The U.S. Supreme Court

The U.S. Supreme Court consists of the Chief Justice and eight associate justices. At its discretion, and within guidelines established by Congress, the Supreme Court each year hears a limited number of cases, which often begin in the federal or state courts, usually involving important interpretations of the U.S. Constitution or federal law.

The courts tend to follow prior rulings, when available, in the same court and jurisdiction. They may also rely on rulings in other jurisdictions, as well as published rulings and other documents issued or provided by the IRS. Cost segregation practitioners should be knowledgeable of prior, related rulings and the weight they carry as precedential evidence.

Revenue Procedures

Along with tax law, the IRS's revenue procedures provide explanations of deductions and how to calculate them in classifying assets for depreciation in cost segregation studies. This guidance includes IRS interpretation of the law and methods for handling certain transactions. The main Revenue Procedure relating to depreciation is 87-56, also reprinted in IRS Publication 946, *How to Depreciate Property.* The publication outlines who can take deductions and when, and how to qualify property for shorter life depreciation. This procedure also includes calculation guidance and definitions, as well as reporting information. Publication 946 is updated frequently to reflect changes in the depreciation rules and asset classifications.

Letter Rulings

When the IRS finds that a particular issue affects numerous taxpayers, it may issue a revenue ruling or interpretation. Both the IRS and the taxpayers are bound to follow revenue rulings. Taxpayers may rely on these rulings only if their tax situation applies to the situation described in the ruling or interpretation—referred to as the "facts and circumstances" rule.

Revenue rulings are published weekly in the *Internal Revenue Bulletin,* and semi-annually in the *Cumulative Bulletin* (both available on the IRS Web site http://www.irs.gov). Cost segregation practitioners should have access to these rulings for use in supporting tax positions taken, especially those applicable to their clients, and in some cases to be included in cost segregation study documentation.

Private Letter Rulings (PLRs) are issued directly to a taxpayer in response to a request for guidance based on particular fact patterns. While these types of rulings are not generally considered precedents, taxpayers may consult them for insight to form opinions based on similar "facts and circumstances." PLRs are made available after all private information is removed, and the ruling is presented in a generic way. *(PLRs are also available on the Internet at http://www.irs.gov as part of the Internal Revenue Bulletin.)*

Internal memoranda issued by the IRS, sometimes called *Technical Advice Memoranda* (TAMs), are available to the public as well on the IRS Web site. TAMs address questions raised during the examination of returns or during other proceedings, and during day-to-day IRS operations. TAMs include requests for clarification from a higher authority within the IRS. Each memorandum is named according to the person who originated the question or raised the issue. Types of TAMs include:

- Field Service Advice (FSA)
- General Counsel Memoranda (GCM)
- Chief Counsel Memoranda (CCM)

To ensure the quality of a cost segregation study, it is important to stay abreast of depreciation-related rulings issued through Letter Rulings and TAMs.

The IRS also publishes for internal use *Audit Techniques Guides,* which are designed to provide insight into particular issues related to taxpayer types, industries, and/or practices. These are used by field auditors. Recently, the IRS issued the *Cost Segregation Audit Techniques Guide* for cost segregation studies, available on its Web site (http://www.irs.gov/businesses/article/0,,id=129356,00.html). Industry-specific guidance is expected to follow. *(See this book's Web site for notification of updates, http://www.rsmeans.com/supplement/67345.asp, or check the IRS Web site for new documents.)* Recent internal guidance at the IRS also addresses cost segregation for the restaurant and casino industries. The restaurant industry guidance is reproduced in Appendix B of this book.

Conclusion

Finding specific rulings or cases to support particular positions taken in cost segregation studies can be challenging. Fortunately, there are many types of support in the area of tax law. In addition to the information available from the IRS, there are searchable databases of current and historic rulings from recognized publishers such as Commerce Clearing House (CCH), Research Institute of America (RIA), and the Bureau of National Affairs (BNA).

Chapter 4
Tax Cases, IRS Revenue Rulings, & Revenue Procedures

Although not a new concept in tax accounting, cost segregation has become increasingly popular since the 1997 *Hospital Corporation of America v. Commissioner* Tax Court case. This landmark ruling for cost segregation established the validity of using the former Investment Tax Credit (ITC) regulations *(see Chapter 3, "The Legal Framework for Cost Segregation")* to determine whether property could be considered tangible personal property. The court ruled that, to the extent an item of property would have qualified for the ITC, it could also be properly treated as tangible personal property for the purpose of calculating the depreciation allowable under the MACRS (Modified Accelerated Cost Recovery System). *(See Chapter 5 for more on MACRS.)*

Previously, there were no established guidelines, so practitioners had relied solely on professional judgment. They applied the circumstances of their client's situation to IRS regulations and publications, to produce typically conservative asset classifications, which were subject to IRS scrutiny.

This chapter will note the most important court cases, rulings, and IRS guidance relating to depreciation and classification of property. Most of the cases were decided during the period of time the ITC was in effect, from the early 1960s through its repeal in 1986. The rulings and published IRS guidance relate to both the ITC and depreciation under the ACRS and MACRS systems. Although the ACRS system is no longer in use, the classification issues discussed in that period apply, for the most part, to the current system of MACRS. *(For further discussion of how Tax Court cases and IRS rulings affect cost segregation procedures, see Chapter 3.)*

May Taxpayers Rely on Cost Segregation Studies Prepared by a Third Party?

One of the first issues to face the courts and the IRS in the development of cost segregation was whether taxpayers could rely on a third-party study to establish asset classifications and depreciation. The resulting rulings not only answered this question, but also helped define the qualifications needed by a third party to perform a cost segregation study.

Letter Ruling 7941002, June 25, 1979

This ruling allowed taxpayers to use third-party cost segregation studies for the allocation of certain costs in a new building as either assets that qualify for the ITC (tangible personal property), or costs properly treated as part of structural building components. A cost segregation study, meeting certain requirements, could be used both to identify qualifying assets and to determine the basis of assets for depreciation purposes.

In this particular case, the taxpayer requested that the general contractors (who constructed the property) separate costs into proper categories for purposes of the ITC. The taxpayer then hired a consulting firm to make an independent estimate of the cost of each new building component and other assets in the construction. Essentially, the taxpayer changed the method of deriving costs. The actual overall cost was then allocated among the property items, based on relative estimated costs.

This Letter Ruling also helped define standard qualifications for cost segregation practitioners, finding, "the consultant is a professional engineering and consulting firm competent in design, construction, auditing, and estimating procedures relating to building construction." The ruling listed these steps as performed by a qualified consultant in completing a cost segregation analysis:

1. An on-site inspection of the facility is made to determine its character, examine the processes and functions for which it is intended, and identify the components considered by the consultant to be property qualifying for the Investment Tax Credit.
2. From a complete set of "as-built" plans and specifications, a detailed quantitative takeoff is made of all materials contained in the facility.
3. The in-place cost of each item of property is then computed by applying applicable unit costs against the indicated quantities of materials and necessary labor. The unit costs are established by specialists and are made applicable with respect to time (period of construction) and geography (location of the project). Where possible, the actual installed cost of the items is obtained from the contractor.
4. The facility's components are then listed by contractor trade, subcontract, or specific purchase order.
5. The facility's actual total construction cost is then distributed among the components based on the relative reconstructed costs.

6. The costs of architectural design, engineering, and miscellaneous capitalized costs are then allocated to the components.
7. The property items are separated into two categories: (1) building components and (2) tangible personal property or other tangible property considered to be Section 38 property.

The ruling also gave the taxpayer the responsibility of showing that the make-up of the component accounts and other tangible property accounts reflects the claims on tax returns, and that these accounts, in fact, qualify for the Investment Tax Credit (i.e., are tangible personal property or other property not relating to the operation or maintenance of buildings).

(Sources for this ruling included Revenue Ruling 66-111, 1966—1 C.B. 46 and *Herbert Shainberg v. Commissioner, 33 T.C. 241.)*

Revenue Ruling 66-111, 1966—1 CB 46

This ruling addressed the classification of component costs by qualified personnel competent in design, construction, and auditing. Where property ledgers for separate accounts were maintained, the original owner can use properly allocated costs as a basis for depreciating the components separately, using an appropriate rate of depreciation for each component account.

Loosely applied to the current methods under MACRS, this case established that taxpayers may classify costs in a construction project as either building-related or personal property, provided records of how the values were arrived at are maintained. These values could be used for calculating allowable depreciation deductions.

Are Site Construction Activities Depreciable Assets or Additions to Non-Depreciable Land?

Revenue Ruling 65-265

In this ruling, a taxpayer constructed a processing and storage complex consisting of groups of buildings located to optimize product handling and transportation. Safety requirements necessitated separating storage buildings by at least 70–80 feet. Not more than 16 buildings were permitted in each group, with a separation of at least 500 feet between groups. A system of level roads was required to provide access to all buildings, to be traversed by forklift trucks to transport materials between the processing plant and storage buildings. The cost of grading the land and excavating was substantial, and a large percentage was directly associated with constructing the buildings and roadways.

While raw land is not depreciable, land preparation costs may be depreciable if they are directly associated with the construction of the depreciable building or land improvement.[1] The regulations state that a depreciation allowance does not apply to land, apart from the improvements for physical development added to it.

In this ruling, the cost of general grading not directly associated with the construction of the building was added to the basis of the land. However, excavating, grading, and removal costs directly associated with the construction of buildings and paved roadways are not inextricably associated with the land itself. These costs, since they have a direct association with the planned construction, are considered part of the costs of constructing the buildings and roadways. The costs attributable to excavation, grading, and removing soil necessary for proper building, and for paving the roadways, are part of the cost of those assets and should be included in the depreciable base for the buildings and roadways.

What If Site Construction Creates Additional Buildable Land?

Revenue Ruling 77-220

In this ruling, a taxpayer incurred costs associated with backfilling a lake to create additional land. These expenditures were treated as *non-depreciable* land.

Eastwood Mall Inc., 95-1 USTC ¶50,236 (CA-6)

In this case, the IRS did not dispute the taxpayer's depreciation of costs for excavation (for trenches required for utility and sewer installation, and for the building's foundation). They also allowed depreciation for costs to pave roads and parking lots. However, it was determined that clearing, grubbing, blasting, filling, and grading 100 acres of uneven land into an earthen plateau in order to construct a shopping mall was not depreciable, since these improvements were permanent and would not be re-incurred if the mall building was rebuilt or replaced.

What If the Client Is Not the Owner of the Improvements Added to the Land?

Revenue Ruling 80-93

This ruling established that the cost of utilities installed by a utility company (responsible for maintaining or replacing the assets as needed) added to the value of the land and was not depreciable. The ruling illustrated which land preparation costs could be depreciated by a developer, and which must be included in the basis of the land.

The particular taxpayer owned and operated a mobile home park and rented sites to tenants on a monthly basis. The taxpayer constructed asphalt roadways, concrete sidewalks, laundry facilities, storm sewers, and other assets at each individual site, including a concrete driveway, patio, and mobile home supports. Other land preparation costs were incurred to excavate and backfill for the laundry facilities and storm sewer system. Replacement of the underground components of these facilities owned by the taxpayer also required excavation and backfilling.

At the taxpayer's request, a public utility company installed an underground electrical distribution system instead of a less expensive above-ground system. The taxpayer contracted to pay the additional costs, including the cost of a natural gas distribution system. The utility company would retain full ownership, provide all necessary maintenance, and replace the system as necessary.

The court ruling held that land preparation costs, such as those for necessary clearing, grubbing, cutting, filling, and rough finish grading that would be unaffected by the placement of the components of the mobile park and therefore would not be replaced contemporaneously, are nonrecurring general land improvement costs. These one-time land preparation costs are considered inextricably associated with the land, and, as such, were to be added to the taxpayer's cost basis in the land and were not considered depreciable.

Land preparation so closely associated with assets (such as the storm sewers and laundry facilities) that they would be retired, abandoned, or replaced contemporaneously with the related assets, could be capitalized and depreciated over the estimated useful lives of the assets with which they are associated.

How Should Demolition Costs Be Treated?

Demolition in a construction project can take many forms. The treatment of costs depends on circumstances surrounding the demolition. Consider the following examples.

Revenue Ruling 2000-7

In this ruling, the cost of removing telephone poles to install new ones was treated as retirement of an asset. If removing a depreciable asset occurs in connection with the installation or production of a replacement asset, the ruling established that the costs incurred are not required to be capitalized as part of the replacement asset. Instead, they are treated as a cost of disposing of the old asset. The ruling, however, did not address rehabilitation plans for larger construction projects.

Revenue Ruling 95-27

When a building is demolished to allow for the construction of a new one, IRC Section 280B provides that *"in the case of the demolition of any structure, no deduction otherwise allowable shall be allowed to the owner (or lessee) for any amount expended for the demolition or any loss sustained on account of the demolition. The expenditure or loss shall be treated as property chargeable to the capital account with respect to the land on which the demolished structure was located."*[2] Stated differently, if you demolish an existing building to build its replacement, the cost of the demolition and the basis of the demolished structure become part of the cost of the land. As we know, the land is not depreciable for tax purposes.

For purposes of this limitation, the term *structure* refers to a building as defined in Section 1.48-1(e), including its structural components. When a portion of the building is demolished, a "safe harbor" provision comes into play. If 75% of the building's internal and external structure is left untouched, then the modification to the building is not treated as a demolition for purposes of IRC Section 280B. The cost would then be taken into account under other sections of the Internal Revenue Code (162, 168, 263, and 263 A). These costs become an allocable indirect cost of the project—allocable to the building in most cases unless the facts and circumstances of the construction call for a different type of allocation.

Is There Any One Case That Would Be Considered a "Must-Read"?

Hospital Corporation of America and Subsidiaries v. Commissioner 109 TC 21, July 24, 1997

This case is widely considered the major landmark decision in relation to the practice of cost segregation and should be read by anyone contemplating cost segregation at any level. Not only did this case establish that Investment Tax Credit tests developed under prior law were applicable to issues related to classifications for depreciation purposes, but the court ruling also addressed a myriad of other asset classification questions, including the following:

- Primary and secondary electrical distribution systems
- Branch wiring systems
- Wiring and related property items involving kitchen equipment
- Special electrical equipment
- Laboratory wiring
- Hospital equipment
- Television equipment
- Telephone equipment
- Internal communications systems
- Carpeting
- Vinyl wall and floor coverings
- Kitchen water piping
- Kitchen equipment steam lines
- Kitchen hoods and exhaust systems
- Steam boilers and related accessories
- Special plumbing connections relating to X-ray equipment
- Corridor handrails for patients
- Lighting over hospital beds
- Accordion doors and partitions
- Bathroom accessories and partitions
- Acoustical tile ceilings

The court ruled that most of these assets were to be treated as Section 1245 property, which is depreciable over five years. The case provided a detailed description of the installation of each individual asset, many of which are common to commercial construction. These are summarized on the following pages.

Acoustical Ceilings

Each individual asset poses unique classification questions. Acoustical ceilings, for example, are typically removable without causing damage to the building structure. The taxpayer argued that, accordingly, these ceilings were not structural components, but constitute Section 1245 property, depreciable over a five-year period. The IRS contended that acoustical ceilings are structural components. They stated that the ceilings were virtually identical to those involved in the Metro National Corporation case,[3] in which the court found these materials to be structural components. The same result was reached in another case, *Texas Instruments Incorporated v. Commissioner.*

The taxpayer also argued that the ceilings had greater similarity to bank vault doors or teller windows, in the sense that their removal would not affect the continued operation of the building, although occasional removal for building repairs could affect the continued operation of the business. [In Revenue Ruling 65-79 1965-1 CB 26, the IRS had ruled that these types of assets (vault doors, etc.) were tangible personal property.]

The court pointed out regulations that treat ceilings and ceiling tiles as structural components of buildings. Once an asset is listed in the regulations as being a structural component, it is the taxpayer's responsibility to prove that the asset more closely relates to the operation of machinery or a process related to a trade or business than it does to the operation and maintenance of the building. Examples of assets that relate more to a trade or business include identity symbols or signs, wiring and plumbing connections for a specific machine, display and specialty lighting, and so forth. While these items are installed as an attachment to the building, the building would be able to operate without them.

Electrical Wiring

Several of the assets addressed in this case included electrical components. The primary distribution systems include the main panel boxes, main motor control centers, transformers, secondary distribution channels, and related wiring conduit. The taxpayer contended that the portion of the system that carried electrical loads to particular items of business equipment constituted Section 1245 property within the meaning of IRC Sections 1.48-1(c) and 1.1245-3(b) of the regulations, because that portion of the electrical system does not relate to the operation or maintenance of the building. The IRS countered that those portions of the system were indeed structural components of the buildings to which they relate, because they are inherently permanent and are not readily removable.

The court referred to Morrison, Inc. and Scott Paper Company in reaching its decision: "In Scott Paper Company, we concluded that the components of the taxpayer's primary election system, which supplied the electric needs of process machinery rather than contributing to the overall operation or maintenance of buildings, was not an inherently permanent structure and the taxpayers were entitled to the Investment Tax Credit on these components."

In reaching this conclusion, the court emphasized the language of Section 1.48-1(e) (which defines structural components by listing examples), and referred to the last line of the regulations: "other components relating to the operation or maintenance of a building." The court characterized the phrase as follows: "a descriptive phrase intended to represent the basic test used for identifying structural components."

The preceding elements are examples of items that meet the test as a general rule. Items that occurred in unusual circumstances and do not relate to the operation or maintenance of a building should not be categorized as structural components despite being listed in Section 1.48-1(e).

The court referred to Central Citrus Company 58 TC 365 374 (1972): "We concluded that the property fails to qualify for the credit if the item relates to the overall operation or maintenance of a building rather than being used to aid in the employment of a particular function or particular piece of property."

The court focused on the ultimate use of power in Scott Paper facilities and concluded that, to the extent that the primary electric system carried electrical loads to be used for pulp and paper production processes or other such qualifying uses, the investment credit will be allowed for the primary electric improvements. To the extent that primary electric improvements relate to the overall operation or maintenance of buildings, they are structural components of such buildings, and they cannot qualify as tangible personal property.

Branch electrical wiring, as it related to the following assets, was also discussed:

- Battery packs and chargers for an emergency power generation system
- X-ray film processing equipment
- Illuminated emergency exit signs
- Illuminated front entrance signs
- Medical gas control equipment
- Medical test control alarms
- Synchronously wired clock systems
- Air conditioners in computer rooms
- Sterilization equipment
- Kitchen equipment
- Shop equipment
- Laboratory equipment
- Local power outlets
- Television equipment wiring, conduit, and antennas
- Telephone wiring, floor boxes, and outlet jacks
- Intercom systems
- Dictation systems
- Nurse call systems

Once again, the court referred to Scott Paper and related rulings to determine the classification of these assets based on the ultimate use of power. In IRS Revenue Ruling 66-299, 1966-2 CB, it was concluded that special electrical or plumbing connections that are necessary for and are used directly with a specific piece of machinery or equipment (or between specific items of individual machinery or equipment) are not structural components of the building and qualify as Section 38 property for Investment Tax Credit purposes.

Local power outlets, including receptacles and junction boxes located in laboratories and shop areas, to the extent that they are used to power employees' personally owned equipment or equipment relating to the operation or maintenance of buildings, do not constitute personal property. Rather, they are structural components of buildings. To the extent they are used to power equipment that is not employee-owned or that does not relate to the operation or maintenance of buildings, however, they do constitute personal property. The taxpayer in this case did not provide evidence showing a qualified use for electrical outlets. Accordingly, the electrical outlets, junction boxes, and related wiring were considered structural components of buildings to be depreciated over the same recovery period as the buildings.

Carpeting

The Hospital Corporation of America case addressed carpeting installed in medical facilities during construction using a general-purpose latex adhesive. Most of the carpeting had already been removed in this particular case. The taxpayer argued that the carpeting was easily removed without damaging the floors underneath, and that, if necessary, it could have been reinstalled in other locations with new adhesives. The IRS argued that this installation was more permanent than Revenue Ruling 67-349, where the carpet was fastened with hooks attached to wood strips that were nailed along the walls. In that ruling, the IRS claimed carpeting installed with adhesive is not an integral covering of the floor and is not a structural component. The petitioners had shown that they did not intend for the carpeting to be permanently affixed to the underlying floor, and that much of the carpeting already had been removed. The court pointed to the Senate Finance Committee Report 95-1263, 1978-3 CB, which specifically identified carpeting as tangible personal property.

Which Rulings List the Assets That Qualify for Accelerated Depreciation?

Court cases like *Texas Instruments v. Commissioner* addressed as many as 200 assets. The following ruling focused on the qualification of some of the more common items for the Investment Tax Credit. Appendix B of this book also includes an alphabetical listing of approximately 300 assets, along with the case or ruling in which they are discussed.

Letter Ruling 7102269400A

This ruling request sought advice as to whether certain assets would qualify for the ITC, including:

1. Air handling and safety equipment
2. Water and power supply distribution facilities
3. Individual air conditioning units
4. Unit heaters
5. Loading docks
6. Roadways
7. Outdoor storage pads
8. Sheds, canopies, and indoor storage areas
9. Research roadways
10. Special assets
    a. Rolling doors
    b. Raised floors for computers
    c. Wood block flooring

d. Scales pits
e. Catwalks
f. Retaining walls
g. Stairways

In its response to this request, the IRS concentrated on the differences between the structural components and those items that ordinarily would not be treated as structural components of buildings (since they qualified as Section 38 property and relate to a process or to machinery used in the manufacturing process).

Air Handling and Safety Equipment

Air handling and safety equipment that relates to the overall operation and maintenance of a building is considered a structural component. To the extent that it is required for the operation of an individual item of machinery or equipment in the manufacturing process, however, it would qualify for the Investment Tax Credit.

Each component was addressed in terms of whether it related to the manufacturing process or to the overall operation and maintenance of the building. The unit heaters in individual air conditioning units, for example, were treated as supplemental to the building's air conditioning and heating system unless they related to a specific process that required a specific temperature and humidity level. As a result, they were considered structural components and not qualifying property.

Loading Docks

Classification of loading docks received a split decision: "Many of the loading docks are fully enclosed areas of a building, while other docks are additions to or extensions of the building. In instances where a dock extends from an area outside the building to an area inside the building, that portion of the loading dock outside the building up to the outermost portion of the building (generally the outer door) qualifies as other tangible property, while the remainder of the dock within the boundaries of the building is a structural component of the building and does not qualify for this Section 38 credit (ITC)." In other rulings, however, loading docks and related equipment have been treated differently under various circumstances.

Sheds, Canopies, and Storage

In this case, sheds, canopies, and indoor storage areas received the following response: "Sheds and canopy storage areas attached to a building that are appendages to the building are not structural components of the buildings. They are inherently permanent structures that may qualify as Section 38 property, depending on the use to which the property beneath the sheds and canopies is put."

Raised Flooring

Raised flooring used for computer installations (built-up false floors) qualified as tangible personal property within Section 1.48-1(c). Flooring originally constructed to service the normal building needs or accommodate computer service installations did not qualify as Section 38 property.

Catwalks

Generally, catwalks qualify as Section 38 property, since they relate to items of machinery and equipment. Catwalks used as stairs are, in effect, stairways used for employees' benefit, and do not qualify as Section 38 property.

Retaining Walls

Retaining walls used to contain raw materials necessary as an integral part of the production activity qualified as tangible property.

Do the Taxpayer's Construction or Design Practices Affect Classification Issues?

Metro National Corporation v. Commissioner TC Memo 1987-38

The Metro National Corporation case, a Tax Court memo decision filed on January 20, 1987, is a good illustration of how a taxpayer could design a building to maximize depreciation deductions, and how taxpayer practices affect the treatment of an asset for tax purposes. The case also addresses other assets that may be of interest.

Metro was formed by Joe Johnson, who started his career as a plumbing subcontractor in residential sections of Houston. In the mid 1950s, he began developing apartment complexes in the suburbs, and later built the largest enclosed shopping mall in the Southwest. In the late 1960s, Metro constructed Memorial City Professional building, referred to as "Pro I" and "Memorial City General Hospital." As the population increased, the professional building was remodeled and expanded over a period of years.

The interior walls of Pro I were constructed with 2 x 4 wood studs covered by gypsum board. These materials were neither removable nor reusable, and thus did not provide any flexibility to accommodate changes in the interior space.

In 1979 and 1980, Metro began design and construction work on two new buildings: Memorial Professional Building II (referred to as "Pro II")—a three-story building with a center atrium extending to the roof, located across the street from the hospital, and Springs Shadows ("The Glen"), a psychiatric treatment facility. Both buildings were completed in 1981. Based on Metro's prior experience with medical facilities space, these buildings were designed to provide maximum flexibility in the configuration of office and other interior space.

These buildings were designed on modular plans that provided easy access to plumbing, heating, air conditioning, ventilation, and electricity. Pro II was designed with 25' modular centers that provided support for the building structure. The heating, ventilating, air conditioning, and electrical lines were run through chases located in the support columns, an approach that facilitates access and minimizes restrictions in interior space configurations.

Pro II, designed for lease to doctors and dentists, was not pre-leased before it was built. The architectural plans for the building did not include the partitions. The interior space was typically not completed until the tenant leased the space and designated his or her floor plan. At the time the space was leased, Metro itself, a general contractor, or a subcontractor built out the space to the tenants' specifications. Bookcases, storage, and cabinetry were specified by the tenants and constructed in Metro's workshop. Partitions, window coverings, floor coverings, ceilings, and lighting were installed as designated by the tenant. Partitions, ceilings, and lighting in The Glen are similar to those used in Pro II.

This case was heard to determine whether Metro was entitled to an Investment Tax Credit with respect to the following assets:

1. Partitions, including gypsum wall, storefront, and toilet partitions
2. False ceilings, grids, and lay-in light fixtures
3. Exterior security lighting, grow lights, and accent decorative lighting
4. Cabinets, laminated plastics, and cabinet hardware
5. Sprinkler heads

The court referred to Congressional intent found in committee reports to explain the general legal principles that applied. In a U.S. House of Representatives committee report, the court noted,

> *"Section 48(a)(1)(A) defines Section 38 property to include, among other items, tangible personal property, but it does not define this elusive term. The income tax regulations explain 'the fact that under local law property is held to be personal property or tangible property does not control a decision whether the property constitutes tangible personal property. Rather, the term is giving a special meaning tailored to the purposes of the investment tax credit provisions.'"*[4]

The U.S. Congress intended that the tangible personal property definition should be flexible or broad enough to include any assets that support a particular business function.

In a U.S. Senate report, the U.S. Congress stated that...

> "...tangible personal property is not intended to be defined narrowly here, nor to follow the rules of state law. It is intended that assets accessory to a business, such as..."
>
> - Grocery store counters
> - Printing presses
> - Individual air conditioner units, etc.

...even though fixtures, under local law, qualify for the credit.[5]

In a later report clarifying Congressional intent, the committee stated that tangible personal property that already qualifies for the credit includes:

- Special lighting–exterior building illumination (does not include parking lot lighting)
- False balconies
- Exterior ornamentation
- Movable partitions
- Removable partitions

Because these items have no more than an incidental relationship to the operation or maintenance of buildings, they are not structural components.

Partitions

With respect to partitions, the IRS argued that the manner of attachment to the land and how permanently the property is designed to remain in place *(permanency)*, described in Revenue Ruling 75-178, is the controlling issue in deciding whether an item is a structural component of a building. The taxpayer argued that the emphasis should be placed on movability, not the manner of attachment, and argued that the IRS's stress on manner of attachment and permanency calls for a rule based on state law personal versus real property concepts, while the regulation expressly stated that state law is not controlling.

The court responded that items listed in the regulations are illustrative of structural components—as a general rule. It pointed to the final element in the regulation, "and other components relating to the operation and maintenance of a building." The court discussed its findings in the Scott Paper case, where it stated:

> *Although we give great weight to the respondent's regulations, it would be improper to read the words "electric wiring" in a vacuum and conclude that those electric cables, therefore, must be structural components. Rather the effect of the final element of the same paragraph which reads "and other components relating to the operation and maintenance of a building" must be taken into account. That final element functions as a descriptive phrase intended to present the basic test used for identifying structural components.*
>
> *The preceding elements are examples of items (those listed in the regulations) that meet the test as a general rule. Items that occur in unusual circumstance and do not relate to the operation and maintenance of a building should not be structural components despite being listed in the regulations. In Section 1.48-2(e)(2), the critical test is whether the assets relate to the overall operation and maintenance of a building.*[6]
>
> *The gypsum board partitions in the Pro II and The Glen are not weight-bearing and are comprised of a sheet metal frame, screwed together. The frame's bottom channel is attached to the concrete floor with masonry nails. The top channel is attached with a brace and screws at intervals of 4' to 8'. Sheet metal studs are screwed to the floor channels and the top channels. Standard-sized gypsum board panels are screwed to the metal frames, taped at the seams, smoothed with filler, and sanded and painted or covered by wallpaper or paneling. The taxpayer presented the fact that the partitions were economically moved and reused. They also showed that this process of reusing the materials resulted in a savings when compared to using new materials.*

The court again stated that Congressional intent was not to prohibit all partitions from eligibility for the ITC, but that the credit should be allowed for movable and removable partitions that have only an incidental relationship to the operation and maintenance of the building. The facts in this case show that these partitions meet the criteria. The court also reviewed the tests developed in Whiteco Industries to answer the question of permanency.

The court also mentions that four cases deal with partitions. The courts allowed the credit in *King Radio Corp. Incorporated v. United States* and *Minot Federal Savings and Loan Association v. United States.* They did not allow the credit in *Mallinckrodt, Incorporated v. Commissioner* and *Dixie Manor Incorporated v. United States.*

Storefront Partitions: Storefront partitions were used in two different applications. They were constructed in a manner similar to that used for the gypsum wall panels discussed earlier in this case, with the doors and panels comprised of glass supported by anodized aluminum framing. The court identified a difference here between those storefronts facing the atrium and those in the interior corridors of the building. The storefronts facing the atrium were determined to be structural components of buildings because the building could not operate as leased space for offices of doctors and dentists without them.

The walls were also considered part of the structural design of the building. Although they could be reconfigured, they could not be removed or eliminated entirely. This is different from the storefronts in the interior corridors, which could be moved, removed, or eliminated to meet the needs of individual tenants.

The storefronts facing the interior corridors were eligible for the Investment Tax Credit because they are of the type in U.S. Senate Report No. 95-1263 (1978), 1978-3. They are permanently installed, but may be moved at the request of the tenants (as illustrated by the elimination of two interior corridors at tenant request). The removal of these partitions does not adversely affect the utility of the building or cause any structural damage.

Toilet Partitions: Toilet partitions were also addressed in this case. "We do not think these partitions qualify for the credit, because they are structural components of the building. The location of the toilets, (which are) essential to the operation of the building, is controlled by the plumbing facilities which are quite obviously structural components of buildings. As long as there are toilets, partitions will be needed; they are thus related to the operation and maintenance of the building as rental space."

False Ceilings and Lay-in Lighting

False ceilings and light fixtures are typical in most office buildings. They are usually 2' x 4' tiles and fixtures installed in a grid system. The taxpayer felt that these items were temporary because they could be replaced with a "true ceiling" at the request of a tenant. The court again stressed that the ability to move an asset is not the only factor to consider. The "movability of the system facilitates normal maintenance above the ceiling, such as adjusting the dampers in the air conditioning ducts, inspecting electrical wiring, adding communication wiring and speakers where needed, adjusting and repairing thermostats, and making routine fire prevention and building inspections. It does not show that they are not structural components of buildings."

The false ceilings and lay-in light fixtures differ somewhat from the gypsum wall partitions, which were planned—from the time the building was designed—to be moved and removed to accommodate the needs of tenants. A tenant's entire leased space could be used without partitions, or the space could be divided into small cubicles. In contrast, ceilings and lighting systems are essential to the operation of the building.

Security and Decorative Lighting

These three types of lighting were in dispute:

1. Accent security lights
2. Grow lights
3. Outside decorative lighting

The accent security lighting described in this case is a typical exterior installation of flood lights. Their removal does not damage the building. The grow lights are used only to light a plant or decorative feature. They, too, are easily removed without damage. Other outdoor lighting is part of a landscape design for ornamental gardens and other features. Outdoor lighting is described in the case as a pedestal light with a frame screwed to the concrete foundation.

The court pointed to U.S. Senate Report No. 95-1263 (1978) and the definition of special lighting in reaching the conclusion that these lighting installations qualified for the Investment Tax Credit. Keep in mind that if the asset would have qualified for the Investment Tax Credit, then that item by definition is tangible personal property.

Cabinets

The cabinets were prefabricated to meet the tenants' needs. The installation included attaching the cabinets to the partitions, to one another, or set in place as free-standing. The cabinets are movable and were, in fact, moved as a common occurrence. Even those cabinets including plumbing could be removed without causing damage to the cabinets or the surrounding area.

The cabinets here were accessory to a business operation within the meaning of U.S. Senate Report No. 1881, and they are not structural components related to the operation or maintenance of buildings. The IRS conceded that the free-standing cabinets were eligible for the credit, and the court did not recognize the difference between free-standing and mounted cabinets, and therefore accepted both as eligible.

Sprinkler Heads

The sprinkler heads were installed on the outdoor irrigation system. The taxpayer unsuccessfully argued that because the sprinkler heads were removable, they qualified for the credit. The court stated that the sprinkler heads are inseparable from and give utility to the underground pipes. The underground pipes are an inherently permanent structure and do not qualify for the Investment Tax Credit.

It is important to note that although the sprinkler heads are not tangible personal property, they also are not structural components of buildings. These types of inherently permanent structures are "land improvements" under current depreciation guidelines and as such are 15-year property for this taxpayer.

Other Cases Worth Reviewing

For tests used to determine personal versus real property, see the discussion on *Whiteco Industries v. the Commissioner* in Chapter 5, on classification of assets. Additional noteworthy cases are listed below and summarized on this book's Web site, **http://www.rsmeans.com/supplement/67345.asp.**

- Albertsons v. Commissioner 94-1 USTC 50,016
- Boddie – Noell Enterprises v. United States 96-2 USTC 50,627
- Central Citrus Company v. Commissioner 58 TC 365
- Circle K Corporation v. Commissioner TC Memo 1982-298
- Consolidated Freightways v. United States 74 TC 768
- Dixie Manor v. United States 79-2 USTC 9469

- Duaine v. Commissioner TC Memo 1985-39
- Grinald v. Commissioner TC Memo 1993-66
- Illinois Cereal Mills v. Commissioner TC Memo 1983-469
- King Radio Corp v. Commissioner 486 F2d 1091
- La Petite Academy v. United States 95-1 USTC 50,193
- LL Bean v. Commissioner TC Memo 1997-175
- Mallinckrodt Inc v. Commissioner TC Memo 1984-532
- McManus v. United States 700 F Supp 994
- Minot Federal Savings & Loan v. United States 313 F Supp 294
- Morrison v. Commissioner TC Memo 1986-129
- Munford, Incorporated v. Commissioner 87 TC 463
- Piggly Wiggly v. Commissioner 84 TC 739
- Publix Supermarkets Incorporated v. United States 92-1 USTC 50,240
- Scott Paper v. Commissioner 74 TC 137
- Shoneys South Inc. v. Commissioner 1984-413
- Texas Instruments v. Commissioner TC Memo 1992-306
- Walgreen Co & Subsidiaries v. Commissioner 103TC 582
- Westroads Incorporated v. Commissioner 69 TC 682

Conclusion

This chapter is intended to summarize some of the cases that could have an impact on classification determinations in cost segregation. It is important in developing a cost segregation practice to accumulate and become familiar with these and other relevant cases and rulings. They should be the starting point for cost segregation study research, not the endpoint.

Any tax classifications that may have an associated audit risk should be reviewed by a tax professional and fully disclosed to a client. The one constant is that taxpayers who are armed with a "quality cost segregation study" can use it as evidence to support classifications and can successfully defend challenges by the IRS.

[1] IRC Regulations Section 1.167 (a)-2.

[2] IRC Section 280B(a).

[3] TC Memo 1987-38.

[4] U.S. House of Representatives Report No. 1447.

[5] U.S. Senate Report No. 1881.

[6] *Scott Paper v. Commissioner* 74 TC 137.

Chapter 5 Classification of Property

Proper classification of property is one of the key elements of cost segregation. To maximize legitimate depreciation deductions, property must be classified into categories qualifying for the optimum rates of depreciation. Cost segregation specialists should understand the different types of classifications and how they are determined. Generally speaking, property is classified into one of five categories:

1. Non-depreciable land
2. Non-residential real property
3. Residential rental property
4. Land improvements (depreciable)
5. Tangible personal property

Land

Land itself is not subject to an allowance for depreciation, though improvements to it sometimes are. The first cost incurred is usually for acquisition, which may include the cost of demolishing any existing structures on the land in preparation for its intended use. Other costs that may be added to the purchase price in order to arrive at the basis of land costs for tax purposes include work that is performed in order to prepare for the construction of a new building. Achieving a level grade, removing trees and stumps, filling in wetland areas, cutting into a hillside, and similar work is considered part of the cost of the land because it adds to the value of the land itself. The cost of bringing utilities to a parcel of land for the first time must also be added to the basis of land costs, although improvements to existing utilities need not be. *(See the discussion on IRS Revenue Ruling 65-265 in Chapter 4.)*

Sections 1245, 1250, & 38 Property

Non-depreciable land, non-residential real property, and residential rental property are commonly referred to as *Section 1250 property.* Tangible personal property has been called both *Section 1245* and *Section 38 property.* The holding in *Hospital Corporation of America v. Commissioner 109 TC 21* made these terms interchangeable.

The crucial element in cost segregation practice is identifying property that may be classified as tangible personal property (items other than buildings and their structural components) or as land improvements (permanent improvements to land *other than* buildings and their structural components). Tangible personal property is depreciated based on the type of asset involved and/or the trade or business activities in which the asset is used.

Figures 5.1 and 5.2 at the end of the chapter, from IRS Revenue Procedure 87-56, outline class lives and recovery periods.

Hospital Corporation of America v. Commissioner 109 TC 21, a landmark case for cost segregation, resulted in the Tax Court's decision stating that tests developed to ascertain whether property that constituted tangible personal property (for the purposes of the Investment Tax Credit, Section 38 property) should be used to make property classifications for the purposes of MACRS depreciation (Section 1245 property). The court stated:

> *"Accordingly, we conclude that, to the extent a disputed property item would have qualified as tangible personal property for Investment Tax Credit, that property also will qualify as tangible personal property for purposes of ACRS and MACRS."*

While many practitioners had long followed this principle, this was the first time it was stated in an authoritative case or pronouncement. *Tangible personal property* does not include permanent improvements to land, such as buildings and their structural components, which are depreciated over 27.5 or 39 years, if their classification is residential rental or non-residential real property. Permanent improvements to land that are not buildings or structural components are referred to as *land improvements*, which are depreciated over 15 years if they are in Table B-1, asset class 00.3, of Revenue Procedure 87-56 and are not included in the classifications for specific activities. Land improvements included in other classifications (specific activities) have the class life assigned to that asset class. *(See asset class 33.4 on Table B-2 of Revenue Procedure 87-56, Figure 5.2.)*

For example, manufacturing of primary steel mill products includes related land improvements, which are excluded from asset class 00.3 and included in asset class 33.4 (which assigns a "class life" of 15 years). Property with a class life of 10 or more years and fewer than 16 years has a recovery period of 7 years ("7-year property").

Following is a list of those assets considered land improvements according to Revenue Procedure 87-56:

- Sidewalks
- Roads
- Canals
- Waterways
- Drainage facilities

- Sewers (not included in asset class 51)
- Wharves and docks
- Bridges
- Fences
- Landscaping shrubbery
- Radio and television transmitting towers
- Utility improvements (other than initial improvements)

Revenue Procedure 87-56 is not all-inclusive, but lists several examples. Other assets considered land improvements include:

- Retaining walls
- Parking lot pavements
- Lighting poles
- Flagpoles
- Lawn sprinkler systems
- Loading dock ramps
- Fire hydrants and related plumbing
- Concrete dumpster pads
- Dumpster enclosures
- Swing sets and playground equipment

Tangible Personal Property

According to Section 1.48-1(c), *tangible personal property* consists of all tangible property except land and inherently permanent improvements to land, such as buildings and their structural components. Practitioners should consult Section 1.48-1(e)(2) of the regulations for the definition of *structural components,* which include such parts of buildings as:

- Walls
- Floors
- Ceilings
- Partitions
- Permanent covering of walls, partitions, floors, and ceilings (such as paneling or tiling)
- Windows and doors
- All components of HVAC systems (including motors, compressors, pipes, and ducts)
- Plumbing and plumbing fixtures (such as sinks or bathtubs)
- Electric wiring and lighting fixtures
- Chimneys
- Stairs, elevators, and escalators
- Sprinkler systems

- Fire escapes
- Other components relating to the operation and maintenance of buildings

According to the regulations, "the term 'structural components' does not include machinery the sole justification for which is meeting temperature or humidity requirements essential to the operation of other machinery or the processing of materials or foodstuffs." This is often referred to as the "sole justification test."

Essentially, if an item is an inherent part of a building, it cannot be considered tangible personal property. The regulations provide several examples. Let's look at some other factors to consider when making this determination.

Congress intended that the items listed in Section 1.48-1 of the regulations as structural components (inherently permanent parts of most structures) be treated as part of the building structure as a general rule, not as business-related components. Where it can be shown that an item relates more to a process or the operation of machinery and equipment necessary to a particular business rather than to the operation or maintenance of a building in general, that item can be properly treated as tangible personal property. Examples in court cases and rulings in this area include special or raised flooring related to electrical wiring installation for machinery and computers.

While floors are included in the regulations as structural components, they are treated as personal property when it can be shown that they are more closely related to the installation of a business-related item of machinery. Specialized air handling systems in computer rooms are another example. While HVAC systems are generally included in the definition of a building's structural components, those that meet temperature or humidity requirements are specifically excluded by the regulations from the definition of structural components because they are business-related.

The courts have also used the "appearance" and "function" tests to determine whether an item of property is structural and inherent to the building itself or a business-related component. In the appearance test, if a structure "looks like a building," then it may be considered a building. What does a building look like? Buildings are defined generally as any structure or edifice enclosing a space within its walls and usually covered by a roof that provides the following types of space:

- Shelter
- Housing
- Working area
- Office
- Parking area
- Display or sales area

Structures that serve the above activities are considered to meet the function test by providing space for those activities. Some examples from the regulations are:

- Apartment houses
- Factory buildings
- Office buildings
- Warehouses
- Barns
- Garages

- Railway stations
- Bus stations
- Stores

Structures that are so closely related to an item of machinery that they would be replaced if the underlying equipment were to be replaced would not be defined as a building or a structural component. Examples include special design to provide for stress or other special demands of such property. The structure should also clearly not be able to be used, economically, for other purposes. Examples include:

- Oil and gas storage tanks
- Grain storage bins
- Silos
- Fractionating towers
- Blast furnaces
- Oxygenating furnaces
- Coke ovens
- Brick kilns
- Coal tipples

These special-use structures are sometimes given their own asset classification or are included in the trade or business classification provided in Revenue Procedure 87-56.

Determining Asset Classification

To determine the classification of an asset, consideration must be given to an item's degree of permanence. Section 1245 property does not apply to land or permanent improvements to land, including buildings or other inherently permanent structures. Whiteco Industries, a landmark 1975 Tax Court case, is referred to time and again in other rulings and cases where the court or the IRS is faced with determining whether an item qualifies as property for purposes of the Investment Tax Credit. Consider the following:

> *White Advertising Company (Whiteco Industries) was in the business of providing billboards for advertising. Their billboards and other signs, attached to the ground, were simple construction made out of common lumber materials. Some, however, involved more sturdy steel and concrete construction. While many signs were located on company-owned land, most were on land leased from private individuals or government agencies along highways. Some components of the signs were considered temporary, because they were often changed, and some of those materials were re-used for other signs. Other sign components had the appearance of being more permanent, especially when the signs were installed on the ground with concrete bases. The Tax Court had to decide whether these items qualified for the ITC—in other words, whether they were inherently permanent structures.*

The Tax Court reviewed the case history and established what is referred to as permanency tests, *comprising the following six questions:*

1. *Is the property capable of being moved, and has it, in fact, been moved?*
2. *Is the property designed or constructed to remain permanently in place?*
3. *Are there circumstances which tend to show the expected or intended length of affixation (i.e., are there circumstances which show that the property may or will have to be moved)?*
4. *How big of a job is removal of the property, and how time-consuming is it? Is it readily movable?*
5. *How much damage will the property sustain on its removal?*
6. *What is the manner of affixation of the property to the land?*

The taxpayers had to prove its position before the court. The proof consisted of a description of the business and of the signs and their construction methods. Whiteco had to show that they considered the signs temporary based on actual data regarding useful life and past practices. They described instances where they had removed these assets in the past, and that they were able to reuse some of the materials in other locations. No one item by itself (immovability) proves an item is not "inherently permanent." Conversely, no one item (permanency) proves that it is.

The court found the taxpayers' advertising signs to be tangible personal property. The most important result of this case is the development of the tests to determine whether an asset is inherently permanent. Because this case is referenced so often, it's an important one to be familiar with in determining whether an item is qualifying property for purposes of the Investment Tax Credit.

The largest impact items in terms of accelerated depreciation are land improvements, which are easily identifiable, for the most part, on the site plans and/or during a site visit. Roadways and retention and drainage facilities are among the major items. Land improvements, in some cost segregation studies, can account for half of the segregated costs that are reclassified for faster depreciation. Some practitioners refer to these assets as the "low hanging fruit," or the easiest to identify of the significant depreciation items that can be accelerated.

Corporate Image-Related Investments

Some construction projects are geared more toward creating an image for the business that operates the facility than operation of business-related machinery. In some cases, the purpose is to establish a recurring corporate design element, such as a distinct building design used in fast-food industry chains. Another image-related improvement would be custom-designed, intricate features applied to the office of an architect, or oak or mahogany woodwork used to establish an image for a high-end sales or a corporate attorney's office.

These are all features within a project that may be able to be segregated for depreciation purposes. To the extent that they can be shown to have only a business-related decorative purpose, they can be classified as personal property, provided they are not "structural" components of a building.

Asset Classification Table

There are literally thousands of components within a construction project that analysts will encounter as they perform cost segregation studies. Each component, or asset, will need to be classified for various taxpayers under different sets of facts and circumstances. Since a taxpayer's own "facts and circumstances" will control classification issues, cost segregation analysts will periodically need a resource to consult regarding decisions for taxpayers who are similarly situated.

The asset classification table in Figure 5.3 at the end of the chapter is a good starting point for this research. It is composed of an alphabetical listing of approximately 300 assets and the relevant cases or rulings that established their classification. Users should take care to read the case, ruling, or revenue procedure before relying on positions developed therein. Rulings or items that, by IRS standards, do not rise to the level of precedent should not be relied on as support for return positions, but can be used in developing a well-reasoned position based on the taxpayer's particular facts and circumstances.

Listed for each asset are:

- The case or ruling of origin
- The relevant citation
- The circuit of the federal court system (if relevant)
- The trade or business of the taxpayer
- The classification assigned to the asset in the given ruling

This listing can be modified with items experienced while performing cost segregation services and will limit the extent of research needed—making better use of the cost segregation analyst's resources.

American Jobs Creation Act of 2004

(Enacted Oct. 22, 2004)

There have not been significant changes to the depreciable lives of non-residential buildings since the 1993 Tax Act, which extended the life of these buildings to 39 years (up from 31.5 years). However, the American Jobs Creation Act of 2004 introduced some new classifications for non-residential buildings. These are "Qualified Restaurant Property" or "Qualified Leasehold Improvement Property." This Act treats the new classifications as 15-year property.

Qualified Restaurant Property (QRP)

QRP was added by the American Jobs Creation Act of 2004 in Section 168(e)(7). To meet the definition of QRP, the improvements must be placed in service after the effective date of October 22, 2004 and before January 1, 2006. QRP is defined as "any" Section 1250 property that is an improvement to a restaurant building, performed more than three years after the building was first placed in service. More than 50% of the building's square footage must be devoted to the preparation of, and seating for, on-premises consumption of prepared meals.

Property that meets the definition of QRP is depreciated using the straight-line method[1] over 15 years.[2] The half-year convention appears to comply unless the mid-quarter convention applies.[3] Prior to the changes made by the 2004 CCT, this property would have been depreciated over 39 years using the straight-line method and a mid-month convention.

Qualified Leasehold Improvement Property (QLIP)

This new classification added by the Act in Section 168(e)(6) is generally given the same definition as "Qualified Leasehold Improvement Property" in Section 168(k) (relating to property qualifying for the Additional First Year or Bonus depreciation). The property must be placed in service after October 22, 2004 and before January 1, 2006. Property that meets the definition of QLIP is depreciated using the straight-line method[4] over 15 years.[5] Again, the half-year convention appears to be used unless the mid-quarter convention applies. Like the Qualified Restaurant Property above, the Qualified Leasehold Improvement Property would previously have been depreciated over 39 years using the straight-line method and a mid-month convention.

Another limitation placed on this type of property is that its treatment as 15-year property does not survive transfers by sale to another party. The property must be held by the person or business enterprise which placed the improvements in service.[6]

These additions to the depreciation scheme were signed as this book went to press. Practitioners should check the IRS' Web site for additional information relating to these new developments. The IRS is certain to provide additional information, as well as revisions to the instructions for Form 4562 in time for the 2004 tax filing season.[7]

Conclusion

The classification of assets as *land, building components, land improvements,* or *tangible personal property* will affect the timing of the depreciation deductions allowed to taxpayers. Personal property and land improvement classifications are preferable to building and non-depreciable land classifications because they help achieve the goal of accelerating depreciation deductions and creating valuable tax deferrals.

The ability to segregate items from one owner to the next will hinge on that business' practices and the particular facts and circumstances. Sometimes the only way to arrive at a particular asset's classification may be to put together a well-thought-out position based on a solid foundation of tax principles that can be supported by case law, regulations, or other IRS pronouncements or rulings in the area.

[1] Section 168 (b)(3)(H)

[2] Section 168 (e)(3)(E)(v)

[3] The mid-quarter convention applies when more than 40% of the current year additions are placed in service in the fourth quarter of the year.

[4] Section 168 (b)(3)(G)

[5] Section 168 (e)(3)(E)(iv)

[6] Section 168 (e)(6)(B)

[7] The filing season refers generally to the period of time from January to April each year when the majority of income tax returns are prepared and filed (or requests for extensions of time to file are made).

Table B-1. **Table of Class Lives and Recovery Periods**

Asset class	Description of assets included	Recovery Periods (in years) Class Life (in years)	GDS (MACRS)	ADS
	SPECIFIC DEPRECIABLE ASSETS USED IN ALL BUSINESS ACTIVITIES, EXCEPT AS NOTED:			
00.11	**Office Furniture, Fixtures, and Equipment:** Includes furniture and fixtures that are not a structural component of a building. Includes such assets as desks, files, safes, and communications equipment. Does not include communications equipment that is included in other classes.	10	7	10
00.12	**Information Systems:** Includes computers and their peripheral equipment used in administering normal business transactions and the maintenance of business records, their retrieval and analysis. Information systems are defined as: 1) Computers: A computer is a programmable electronically activated device capable of accepting information, applying prescribed processes to the information, and supplying the results of these processes with or without human intervention. It usually consists of a central processing unit containing extensive storage, logic, arithmetic, and control capabilities. Excluded from this category are adding machines, electronic desk calculators, etc., and other equipment described in class 00.13. 2) Peripheral equipment consists of the auxiliary machines which are designed to be placed under control of the central processing unit. Nonlimiting examples are: Card readers, card punches, magnetic tape feeds, high speed printers, optical character readers, tape cassettes, mass storage units, paper tape equipment, keypunches, data entry devices, teleprinters, terminals, tape drives, disc drives, disc files, disc packs, visual image projector tubes, card sorters, plotters, and collators. Peripheral equipment may be used on-line or off-line. Does not incude equipment that is an integral part of other capital equipment that is included in other classes of economic activity, i.e., computers used primarily for process or production control, switching, channeling, and automating distributive trades and services such as point of sale (POS) computer systems. Also, does not include equipment of a kind used primarily for amusement or entertainment of the user.	6	5	5
00.13	**Data Handling Equipment; except Computers:** Includes only typewriters, calculators, adding and accounting machines, copiers, and duplicating equipment.	6	5	6
00.21	**Airplanes (airframes and engines), except those used in commercial or contract carrying of passengers or freight, and all helicopters (airframes and engines)**	6	5	6
00.22	**Automobiles, Taxis**	3	5	5
00.23	**Buses**	9	5	9
00.241	**Light General Purpose Trucks:** Includes trucks for use over the road (actual weight less than 13,000 pounds)	4	5	5
00.242	**Heavy General Purpose Trucks:** Includes heavy general purpose trucks, concrete ready mix-trucks, and ore trucks, for use over the road (actual unloaded weight 13,000 pounds or more)	6	5	6
00.25	**Railroad Cars and Locomotives, except those owned by railroad transportation companies**	15	7	15
00.26	**Tractor Units for Use Over-The-Road**	4	3	4
00.27	**Trailers and Trailer-Mounted Containers**	6	5	6
00.28	**Vessels, Barges, Tugs, and Similar Water Transportation Equipment, except those used in marine construction**	18	10	18
00.3	**Land Improvements:** Includes improvements directly to or added to land, whether such improvements are section 1245 property or section 1250 property, provided such improvements are depreciable. Examples of such assets might include sidewalks, roads, canals, waterways, drainage facilities, sewers (not including municipal sewers in Class 51), wharves and docks, bridges, fences, landscaping shrubbery, or radio and television transmitting towers. Does not include land improvements that are explicitly included in any other class, and buildings and structural components as defined in section 1.48-1(e) of the regulations. Excludes public utility initial clearing and grading land improvements as specified in Rev. Rul. 72-403, 1972-2 C.B. 102.	20	15	20
00.4	**Industrial Steam and Electric Generation and/or Distribution Systems:** Includes assets, whether such assets are section 1245 property or 1250 property, providing such assets are depreciable, used in the production and/or distribution of electricity with rated total capacity in excess of 500 Kilowatts and/or assets used in the production and/or distribution of steam with rated total capacity in excess of 12,500 pounds per hour for use by the taxpayer in its industrial manufacturing process or plant activity and not ordinarily available for sale to others. Does not include buildings and structural components as defined in section 1.48-1(e) of the regulations. Assets used to generate and/or distribute electricity or steam of the type described above, but of lesser rated capacity, are not included, but are included in the appropriate manufacturing equipment classes elsewhere specified. Also includes electric generating and steam distribution assets, which may utilize steam produced by a waste reduction and resource recovery plant, used by the taxpayer in its industrial manufacturing process or plant activity. Steam and chemical recovery boiler systems used for the recovery and regeneration of chemicals used in manufacturing, with rated capacity in excess of that described above, with specifically related distribution and return systems are not included but are included in appropriate manufacturing equipment classes elsewhere specified. An example of an excluded steam and chemical recovery boiler system is that used in the pulp and paper manufacturing equipment classes elsewhere specified. An example of an excluded steam and chemical recovery boiler system is that used in the pulp and paper manufacturing industry.	22	15	22

Figure 5.1

Credit: IRS "Revenue Procedure 87-56"

Table B-2. **Table of Class Lives and Recovery Periods**

Asset class	Description of assets included	Recovery Periods (in years) Class Life (in years)	GDS (MACRS)	ADS
	DEPRECIABLE ASSETS USED IN THE FOLLOWING ACTIVITIES:			
01.1	**Agriculture:** Includes machinery and equipment, grain bins, and fences but no other land improvements, that are used in the production of crops or plants, vines, and trees; livestock; the operation of farm dairies, nurseries, greenhouses, sod farms, mushroom cellars, cranberry bogs, apiaries, and fur farms; the performance of agriculture, animal husbandry, and horticultural services.	10	7	10
01.11	**Cotton Ginning Assets**	12	7	12
01.21	**Cattle, Breeding or Dairy**	7	5	7
01.221	**Any breeding or work horse that is 12 years old or less at the time it is placed in service****	10	7	10
01.222	**Any breeding or work horse that is more than 12 years old at the time it is placed in service****	10	3	10
01.223	**Any race horse that is more than 2 years old at the time it is placed in service****	*	3	12
01.224	**Any horse that is more than 12 years old at the time it is placed in service and that is neither a race horse nor a horse described in class 01.222****	*	3	12
01.225	**Any horse not described in classes 01.221, 01.222, 01.223, or 01.224**	*	7	12
01.23	**Hogs, Breeding**	3	3	3
01.24	**Sheep and Goats, Breeding**	5	5	5
01.3	**Farm buildings except structures included in Class 01.4**	25	20	25
01.4	**Single purpose agricultural or horticultural structures (within the meaning of section 168(i)(13) of the Code)**	15	10***	15
10.0	**Mining:** Includes assets used in the mining and quarrying of metallic and nonmetallic minerals (including sand, gravel, stone, and clay) and the milling, beneficiation and other primary preparation of such materials.	10	7	10
13.0	**Offshore Drilling:** Includes assets used in offshore drilling for oil and gas such as floating, self-propelled and other drilling vessels, barges, platforms, and drilling equipment and support vessels such as tenders, barges, towboats and crewboats. Excludes oil and gas production assets.	7.5	5	7.5
13.1	**Drilling of Oil and Gas Wells:** Includes assets used in the drilling of onshore oil and gas wells and the provision of geophysical and other exploration services; and the provision of such oil and gas field services as chemical treatment, plugging and abandoning of wells and cementing or perforating well casings. Does not include assets used in the performance of any of these activities and services by integrated petroleum and natural gas producers for their own account.	6	5	6
13.2	**Exploration for and Production of Petroleum and Natural Gas Deposits:** Includes assets used by petroleum and natural gas producers for drilling of wells and production of petroleum and natural gas, including gathering pipelines and related storage facilities. Also includes petroleum and natural gas offshore transportation facilities used by producers and others consisting of platforms (other than drilling platforms classified in Class 13.0), compression or pumping equipment, and gathering and transmission lines to the first onshore transshipment facility. The assets used in the first onshore transshipment facility are also included and consist of separation equipment (used for separation of natural gas, liquids, and in Class 49.23), and liquid holding or storage facilities (other than those classified in Class 49.25). Does not include support vessels.	14	7	14
13.3	**Petroleum Refining:** Includes assets used for the distillation, fractionation, and catalytic cracking of crude petroleum into gasoline and its other components.	16	10	16
15.0	**Construction:** Includes assets used in construction by general building, special trade, heavy and marine construction contractors, operative and investment builders, real estate subdividers and developers, and others except railroads.	6	5	6
20.1	**Manufacture of Grain and Grain Mill Products:** Includes assets used in the production of flours, cereals, livestock feeds, and other grain and grain mill products.	17	10	17
20.2	**Manufacture of Sugar and Sugar Products:** Includes assets used in the production of raw sugar, syrup, or finished sugar from sugar cane or sugar beets.	18	10	18
20.3	**Manufacture of Vegetable Oils and Vegetable Oil Products:** Includes assets used in the production of oil from vegetable materials and the manufacture of related vegetable oil products.	18	10	18
20.4	**Manufacture of Other Food and Kindred Products:** Includes assets used in the production of foods and beverages not included in classes 20.1, 20.2 and 20.3.	12	7	12
20.5	**Manufacture of Food and Beverages—Special Handling Devices:** Includes assets defined as specialized materials handling devices such as returnable pallets, palletized containers, and fish processing equipment including boxes, baskets, carts, and flaking trays used in activities as defined in classes 20.1, 20.2, 20.3 and 20.4. Does not include general purpose small tools such as wrenches and drills, both hand and power-driven, and other general purpose equipment such as conveyors, transfer equipment, and materials handling devices.	4	3	4

* Property described in asset classes 01.223, 01.224, and 01.225 are assigned recovery periods but have no class lives.
** A horse is more than 2 (or 12) years old after the day that is 24 (or 144) months after its actual birthdate.
*** 7 if property was placed in service before 1989.

Figure 5.2
Credit: IRS "Revenue Procedure 87-56"

Table B-2. **Table of Class Lives and Recovery Periods**

Asset class	Description of assets included	Recovery Periods (in years) Class Life (in years)	GDS (MACRS)	ADS
21.0	**Manufacture of Tobacco and Tobacco Products:** Includes assets used in the production of cigarettes, cigars, smoking and chewing tobacco, snuff, and other tobacco products.	15	7	15
22.1	**Manufacture of Knitted Goods:** Includes assets used in the production of knitted and netted fabrics and lace. Assets used in yarn preparation, bleaching, dyeing, printing, and other similar finishing processes, texturing, and packaging, are elsewhere classified.	7.5	5	7.5
22.2	**Manufacture of Yarn, Thread, and Woven Fabric:** Includes assets used in the production of spun yarns including the preparing, blending, spinning, and twisting of fibers into yarns and threads, the preparation of yarns such as twisting, warping, and winding, the production of covered elastic yarn and thread, cordage, woven fabric, tire fabric, braided fabric, twisted jute for packaging, mattresses, pads, sheets, and industrial belts, and the processing of textile mill waste to recover fibers, flocks, and shoddies. Assets used to manufacture carpets, man-made fibers, and nonwovens, and assets used in texturing, bleaching, dyeing, printing, and other similar finishing processes, are elsewhere classified.	11	7	11
22.3	**Manufacture of Carpets and Dyeing, Finishing, and Packaging of Textile Products and Manufacture of Medical and Dental Supplies:** Includes assets used in the production of carpets, rugs, mats, woven carpet backing, chenille, and other tufted products, and assets used in the joining together of backing with carpet yarn or fabric. Includes assets used in washing, scouring, bleaching, dyeing, printing, drying, and similar finishing processes applied to textile fabrics, yarns, threads, and other textile goods. Includes assets used in the production and packaging of textile products, other than apparel, by creasing, forming, trimming, cutting, and sewing, such as the preparation of carpet and fabric samples, or similar joining together processes (other than the production of scrim reinforced paper products and laminated paper products) such as the sewing and folding of hosiery and panty hose, and the creasing, folding, trimming, and cutting of fabrics to produce nonwoven products, such as disposable diapers and sanitary products. Also includes assets used in the production of medical and dental supplies other than drugs and medicines. Assets used in the manufacture of nonwoven carpet backing, and hard surface floor covering such as tile, rubber, and cork, are elsewhere classified.	9	5	9
22.4	**Manufacture of Textile Yarns:** Includes assets used in the processing of yarns to impart bulk and/or stretch properties to the yarn. The principal machines involved are falsetwist, draw, beam-to-beam, and stuffer box texturing equipment and related highspeed twisters and winders. Assets, as described above, which are used to further process man-made fibers are elsewhere classified when located in the same plant in an integrated operation with man-made fiber producing assets. Assets used to manufacture man-made fibers and assets used in bleaching, dyeing, printing, and other similar finishing processes, are elsewhere classified.	8	5	8
22.5	**Manufacture of Nonwoven Fabrics:** Includes assets used in the production of nonwoven fabrics, felt goods including felt hats, padding, batting, wadding, oakum, and fillings, from new materials and from textile mill waste. Nonwoven fabrics are defined as fabrics (other than reinforced and laminated composites consisting of nonwovens and other products) manufactured by bonding natural and/or synthetic fibers and/or filaments by means of induced mechanical interlocking, fluid entanglement, chemical adhesion, thermal or solvent reaction, or by combination thereof other than natural hydration bonding as ocurs with natural cellulose fibers. Such means include resin bonding, web bonding, and melt bonding. Specifically includes assets used to make flocked and needle punched products other than carpets and rugs. Assets, as described above, which are used to manufacture nonwovens are elsewhere classified when located in the same plant in an integrated operation with man-made fiber producing assets. Assets used to manufacture man-made fibers and assets used in bleaching, dyeing, printing, and other similar finishing processes, are elsewhere classified.	10	7	10
23.0	**Manufacture of Apparel and Other Finished Products:** Includes assets used in the production of clothing and fabricated textile products by the cutting and sewing of woven fabrics, other textile products, and furs; but does not include assets used in the manufacture of apparel from rubber and leather.	9	5	9
24.1	**Cutting of Timber:** Includes logging machinery and equipment and roadbuilding equipment used by logging and sawmill operators and pulp manufacturers for their own account.	6	5	6
24.2	**Sawing of Dimensional Stock from Logs:** Includes machinery and equipment installed in permanent or well established sawmills.	10	7	10
24.3	**Sawing of Dimensional Stock from Logs:** Includes machinery and equipment in sawmills characterized by temporary foundations and a lack, or minimum amount, of lumberhandling, drying, and residue disposal equipment and facilities.	6	5	6
24.4	**Manufacture of Wood Products, and Furniture:** Includes assets used in the production of plywood, hardboard, flooring, veneers, furniture, and other wood products, including the treatment of poles and timber.	10	7	10
26.1	**Manufacture of Pulp and Paper:** Includes assets for pulp materials handling and storage, pulp mill processing, bleach processing, paper and paperboard manufacturing, and on-line finishing. Includes pollution control assets and all land improvements associated with the factory site or production process such as effluent ponds and canals, provided such improvements are depreciable but does not include buildings and structural components as defined in section 1.48-1(e)(1) of the regulations. Includes steam and chemical recovery boiler systems, with any rated capacity, used for the recovery and regeneration of chemicals used in manufacturing. Does not include assets used either in pulpwood logging, or in the manufacture of hardboard.	13	7	13

Figure 5.2 continued

Credit: IRS "Revenue Procedure 87-56"

Table B-2. **Table of Class Lives and Recovery Periods**

Asset class	Description of assets included	Recovery Periods (in years)		
		Class Life (in years)	GDS (MACRS)	ADS
26.2	**Manufacture of Converted Paper, Paperboard, and Pulp Products:** Includes assets used for modification, or remanufacture of paper and pulp into converted products, such as paper coated off the paper machine, paper bags, paper boxes, cartons and envelopes. Does not include assets used for manufacture of nonwovens that are elsewhere classified.	10	7	10
27.0	**Printing, Publishing, and Allied Industries:** Includes assets used in printing by one or more processes, such as letter-press, lithography, gravure, or screen; the performance of services for the printing trade, such as bookbinding, typesetting, engraving, photo-engraving, and electrotyping; and the publication of newspapers, books, and periodicals.	11	7	11
28.0	**Manufacture of Chemicals and Allied Products:** Includes assets used to manufacture basic organic and inorganic chemicals; chemical products to be used in further manufacture, such as synthetic fibers and plastics materials; and finished chemical products. Includes assets used to further process man-made fibers, to manufacture plastic film, and to manufacture nonwoven fabrics, when such assets are located in the same plant in an integrated operation with chemical products producing assets. Also includes assets used to manufacture photographic supplies, such as film, photographic paper, sensitized photographic paper, and developing chemicals. Includes all land improvements associated with plant site or production processes, such as effluent ponds and canals, provided such land improvements are depreciable but does not include buildings and structural components as defined in section 1.48-1(e) of the regulations. Does not include assets used in the manufacture of finished rubber and plastic products or in the production of natural gas products, butane, propane, and by-products of natural gas production plants.	9.5	5	9.5
30.1	**Manufacture of Rubber Products:** Includes assets used for the production of products from natural, synthetic, or reclaimed rubber, gutta percha, balata, or gutta siak, such as tires, tubes, rubber footwear, mechanical rubber goods, heels and soles, flooring, and rubber sundries; and in the recapping, retreading, and rebuilding of tires.	14	7	14
30.11	**Manufacture of Rubber Products—Special Tools and Devices:** Includes assets defined as special tools, such as jigs, dies, mandrels, molds, lasts, patterns, specialty containers, pallets, shells; and tire molds, and accessory parts such as rings and insert plates used in activities as defined in class 30.1. Does not include tire building drums and accessory parts and general purpose small tools such as wrenches and drills, both power and hand-driven, and other general purpose equipment such as conveyors and transfer equipment.	4	3	4
30.2	**Manufacture of Finished Plastic Products:** Includes assets used in the manufacture of plastics products and the molding of primary plastics for the trade. Does not include assets used in the manufacture of basic plastics materials nor the manufacture of phonograph records.	11	7	11
30.21	**Manufacture of Finished Plastic Products—Special Tools:** Includes assets defined as special tools, such as jigs, dies, fixtures, molds, patterns, gauges, and specialty transfer and shipping devices, used in activities as defined in class 30.2. Special tools are specifically designed for the production or processing of particular parts and have no significant utilitarian value and cannot be adapted to further or different use after changes or improvements are made in the model design of the particular part produced by the special tools. Does not include general purpose small tools such as wrenches and drills, both hand and power-driven, and other general purpose equipment such as conveyors, transfer equipment, and materials handling devices.	3.5	3	3.5
31.0	**Manufacture of Leather and Leather Products:** Includes assets used in the tanning, currying, and finishing of hides and skins; the processing of fur pelts; and the manufacture of finished leather products, such as footwear, belting, apparel, and luggage.	11	7	11
32.1	**Manufacture of Glass Products:** Includes assets used in the production of flat, blown, or pressed products of glass, such as float and window glass, glass containers, glassware and fiberglass. Does not include assets used in the manufacture of lenses.	14	7	14
32.11	**Manufacture of Glass Products—Special Tools:** Includes assets defined as special tools such as molds, patterns, pallets, and specialty transfer and shipping devices such as steel racks to transport automotive glass, used in activities as defined in class 32.1. Special tools are specifically designed for the production or processing of particular parts and have no significant utilitarian value and cannot be adapted to further or different use after changes or improvements are made in the model design of the particular part produced by the special tools. Does not include general purpose small tools such as wrenches and drills, both hand and power-driven, and other general purpose equipment such as conveyors, transfer equipment, and materials handling devices.	2.5	3	2.5
32.2	**Manufacture of Cement:** Includes assets used in the production of cement, but does not include assets used in the manufacture of concrete and concrete products nor in any mining or extraction process.	20	15	20
32.3	**Manufacture of Other Stone and Clay Products:** Includes assets used in the manufacture of products from materials in the form of clay and stone, such as brick, tile, and pipe; pottery and related products, such as vitreous-china, plumbing fixtures, earthenware and ceramic insulating materials; and also includes assets used in manufacture of concrete and concrete products. Does not include assets used in any mining or extraction processes.	15	7	15

Figure 5.2 continued

Credit: IRS "Revenue Procedure 87-56"

Table B-2. **Table of Class Lives and Recovery Periods**

Asset class	Description of assets included	Recovery Periods (in years)		
		Class Life (in years)	GDS (MACRS)	ADS
26.2	**Manufacture of Converted Paper, Paperboard, and Pulp Products:** Includes assets used for modification, or remanufacture of paper and pulp into converted products, such as paper coated off the paper machine, paper bags, paper boxes, cartons and envelopes. Does not include assets used for manufacture of nonwovens that are elsewhere classified.	10	7	10
27.0	**Printing, Publishing, and Allied Industries:** Includes assets used in printing by one or more processes, such as letter-press, lithography, gravure, or screen; the performance of services for the printing trade, such as bookbinding, typesetting, engraving, photo-engraving, and electrotyping; and the publication of newspapers, books, and periodicals.	11	7	11
28.0	**Manufacture of Chemicals and Allied Products:** Includes assets used to manufacture basic organic and inorganic chemicals; chemical products to be used in further manufacture, such as synthetic fibers and plastics materials; and finished chemical products. Includes assets used to further process man-made fibers, to manufacture plastic film, and to manufacture nonwoven fabrics, when such assets are located in the same plant in an integrated operation with chemical products producing assets. Also includes assets used to manufacture photographic supplies, such as film, photographic paper, sensitized photographic paper, and developing chemicals. Includes all land improvements associated with plant site or production processes, such as effluent ponds and canals, provided such land improvements are depreciable but does not include buildings and structural components as defined in section 1.48-1(e) of the regulations. Does not include assets used in the manufacture of finished rubber and plastic products or in the production of natural gas products, butane, propane, and by-products of natural gas production plants.	9.5	5	9.5
30.1	**Manufacture of Rubber Products:** Includes assets used for the production of products from natural, synthetic, or reclaimed rubber, gutta percha, balata, or gutta siak, such as tires, tubes, rubber footwear, mechanical rubber goods, heels and soles, flooring, and rubber sundries; and in the recapping, retreading, and rebuilding of tires.	14	7	14
30.11	**Manufacture of Rubber Products—Special Tools and Devices:** Includes assets defined as special tools, such as jigs, dies, mandrels, molds, lasts, patterns, specialty containers, pallets, shells; and tire molds, and accessory parts such as rings and insert plates used in activities as defined in class 30.1. Does not include tire building drums and accessory parts and general purpose small tools such as wrenches and drills, both power and hand-driven, and other general purpose equipment such as conveyors and transfer equipment.	4	3	4
30.2	**Manufacture of Finished Plastic Products:** Includes assets used in the manufacture of plastics products and the molding of primary plastics for the trade. Does not include assets used in the manufacture of basic plastics materials nor the manufacture of phonograph records.	11	7	11
30.21	**Manufacture of Finished Plastic Products—Special Tools:** Includes assets defined as special tools, such as jigs, dies, fixtures, molds, patterns, gauges, and specialty transfer and shipping devices, used in activities as defined in class 30.2. Special tools are specifically designed for the production or processing of particular parts and have no significant utilitarian value and cannot be adapted to further or different use after changes or improvements are made in the model design of the particular part produced by the special tools. Does not include general purpose small tools such as wrenches and drills, both hand and power-driven, and other general purpose equipment such as conveyors, transfer equipment, and materials handling devices.	3.5	3	3.5
31.0	**Manufacture of Leather and Leather Products:** Includes assets used in the tanning, currying, and finishing of hides and skins; the processing of fur pelts; and the manufacture of finished leather products, such as footwear, belting, apparel, and luggage.	11	7	11
32.1	**Manufacture of Glass Products:** Includes assets used in the production of flat, blown, or pressed products of glass, such as float and window glass, glass containers, glassware and fiberglass. Does not include assets used in the manufacture of lenses.	14	7	14
32.11	**Manufacture of Glass Products—Special Tools:** Includes assets defined as special tools such as molds, patterns, pallets, and specialty transfer and shipping devices such as steel racks to transport automotive glass, used in activities as defined in class 32.1. Special tools are specifically designed for the production or processing of particular parts and have no significant utilitarian value and cannot be adapted to further or different use after changes or improvements are made in the model design of the particular part produced by the special tools. Does not include general purpose small tools such as wrenches and drills, both hand and power-driven, and other general purpose equipment such as conveyors, transfer equipment, and materials handling devices.	2.5	3	2.5
32.2	**Manufacture of Cement:** Includes assets used in the production of cement, but does not include assets used in the manufacture of concrete and concrete products nor in any mining or extraction process.	20	15	20
32.3	**Manufacture of Other Stone and Clay Products:** Includes assets used in the manufacture of products from materials in the form of clay and stone, such as brick, tile, and pipe; pottery and related products, such as vitreous-china, plumbing fixtures, earthenware and ceramic insulating materials; and also includes assets used in manufacture of concrete and concrete products. Does not include assets used in any mining or extraction processes.	15	7	15

Figure 5.2 continued

Credit: IRS "Revenue Procedure 87-56"

Table B-2. **Table of Class Lives and Recovery Periods**

Asset class	Description of assets included	Recovery Periods (in years) Class Life (in years)	GDS (MACRS)	ADS
33.2	**Manufacture of Primary Nonferrous Metals:** Includes assets used in the smelting, refining, and electrolysis of nonferrous metals from ore, pig, or scrap, the rolling, drawing, and alloying of nonferrous metals; the manufacture of castings, forgings, and other basic products of nonferrous metals; and the manufacture of nails, spikes, structural shapes, tubing, wire, and cable.	14	7	14
33.21	**Manufacture of Primary Nonferrous Metals—Special Tools:** Includes assets defined as special tools such as dies, jigs, molds, patterns, fixtures, gauges, and drawings concerning such special tools used in the activities as defined in class 33.2, Manufacture of Primary Nonferrous Metals. Special tools are specifically designed for the production or processing of particular products or parts and have no significant utilitarian value and cannot be adapted to further or different use after changes or improvements are made in the model design of the particular part produced by the special tools. Does not include general purpose small tools such as wrenches and drills, both hand and power-driven, and other general purpose equipment such as conveyors, transfer equipment, and materials handling devices. Rolls, mandrels and refractories are not included in class 33.21 but are included in class 33.2.	6.5	5	6.5
33.3	**Manufacture of Foundry Products:** Includes assets used in the casting of iron and steel, including related operations such as molding and coremaking. Also includes assets used in the finishing of castings and patternmaking when performed at the foundry, all special tools and related land improvements.	14	7	14
33.4	**Manufacture of Primary Steel Mill Products:** Includes assets used in the smelting, reduction, and refining of iron and steel from ore, pig, or scrap; the rolling, drawing and alloying of steel; the manufacture of nails, spikes, structural shapes, tubing, wire, and cable. Includes assets used by steel service centers, ferrous metal forges, and assets used in coke production, regardless of ownership. Also includes related land improvements and all special tools used in the above activities.	15	7	15
34.0	**Manufacture of Fabricated Metal Products:** Includes assets used in the production of metal cans, tinware, fabricated structural metal products, metal stampings, and other ferrous and nonferrous metal and wire products not elsewhere classified. Does not include assets used to manufacture non-electric heating apparatus.	12	7	12
34.01	**Manufacture of Fabricated Metal Products—Special Tools:** Includes assets defined as special tools such as dies, jigs, molds, patterns, fixtures, gauges, and returnable containers and drawings concerning such special tools used in the activities as defined in class 34.0. Special tools are specifically designed for the production or processing of particular machine components, products, or parts, and have no significant utilitarian value and cannot be adapted to further or different use after changes or improvements are made in the model design of the particular part produced by the special tools. Does not include general small tools such as wrenches and drills, both hand and power-driven, and other general purpose equipment such as conveyors, transfer equipment, and materials handling devices.	3	3	3
35.0	**Manufacture of Electrical and Non-Electrical Machinery and Other Mechanical Products:** Includes assets used to manufacture or rebuild finished machinery and equipment and replacement parts thereof such as machine tools, general industrial and special industry machinery, electrical power generation, transmission, and distribution systems, space heating, cooling, and refrigeration systems, commercial and home appliances, farm and garden machinery, construction machinery, mining and oil field machinery, internal combustion engines (except those elsewhere classified), turbines (except those that power airborne vehicles), batteries, lamps and lighting fixtures, carbon and graphite products, and electromechanical and mechanical products including business machines, instruments, watches and clocks, vending and amusement machines, photographic equipment, medical and dental equipment and appliances, and ophthalmic goods. Includes assets used by manufacturers or rebuilders of such finished machinery and equipment in activities elsewhere classified such as the manufacture of castings, forgings, rubber and plastic products, electronic subassemblies or other manufacturing activities if the interim products are used by the same manufacturer primarily in the manufacture, assembly, or rebuilding of such finished machinery and equipment. Does not include assets used in mining, assets used in the manufacture of primary ferrous and nonferrous metals, assets included in class 00.11 through 00.4 and assets elsewhere classified.	10	7	10
36.0	**Manufacture of Electronic Components, Products, and Systems:** Includes assets used in the manufacture of electronic communication, computation, instrumentation and control system, including airborne applications; also includes assets used in the manufacture of electronic products such as frequency and amplitude modulated transmitters and receivers, electronic switching stations, television cameras, video recorders, record players and tape recorders, computers and computer peripheral machines, and electronic instruments, watches, and clocks; also includes assets used in the manufacture of components, provided their primary use is products and systems defined above such as electron tubes, capacitors, coils, resistors, printed circuit substrates, switches, harness cables, lasers, fiber optic devices, and magnetic media devices. Specifically excludes assets used to manufacture electronic products and components, photocopiers, typewriters, postage meters and other electromechanical and mechanical business machines and instruments that are elsewhere classified. Does not include semiconductor manufacturing equipment included in class 36.1.	6	5	6
36.1	**Any Semiconductor Manufacturing Equipment:** Includes equipment used in the manufacturing of semiconductors if the primary use of the semiconductors so produced is in products and systems of the type defined in class 36.0.	5	5	5

Figure 5.2 continued

Credit: IRS "Revenue Procedure 87-56"

Table B-2. **Table of Class Lives and Recovery Periods**

Asset class	Description of assets included	Recovery Periods (in years) Class Life (in years)	GDS (MACRS)	ADS
37.11	**Manufacture of Motor Vehicles:** Includes assets used in the manufacture and assembly of finished automobiles, trucks, trailers, motor homes, and buses. Does not include assets used in mining, printing and publishing, production of primary metals, electricity, or steam, or the manufacture of glass, industrial chemicals, batteries, or rubber products, which are classified elsewhere. Includes assets used in manufacturing activities elsewhere classified other than those excluded above, where such activities are incidental to and an integral part of the manufacture and assembly of finished motor vehicles such as the manufacture of parts and subassemblies of fabricated metal products, electrical equipment, textiles, plastics, leather, and foundry and forging operations. Does not include any assets not classified in manufacturing activity classes, e.g., does not include any assets classified in asset guideline classes 00.11 through 00.4. Activities will be considered incidental to the manufacture and assembly of finished motor vehicles only if 75 percent or more of the value of the products produced under one roof are used for the manufacture and assembly of finished motor vehicles. Parts that are produced as a normal replacement stock complement in connection with the manufacture and assembly of finished motor vehicles are considered used for the manufacture assembly of finished motor vehicles. Does not include assets used in the manufacture of component parts if these assets are used by taxpayers not engaged in the assembly of finished motor vehicles.	12	7	12
37.12	**Manufacture of Motor Vehicles—Special Tools:** Includes assets defined as special tools, such as jigs, dies, fixtures, molds, patterns, gauges, and specialty transfer and shipping devices, owned by manufacturers of finished motor vehicles and used in qualified activities as defined in class 37.11. Special tools are specifically designed for the production or processing of particular motor vehicle components and have no significant utilitarian value, and cannot be adapted to further or different use, after changes or improvements are made in the model design of the particular part produced by the special tools. Does not include general purpose small tools such as wrenches and drills, both hand and powerdriven, and other general purpose equipment such as conveyors, transfer equipment, and materials handling devices.	3	3	3
37.2	**Manufacture of Aerospace Products:** Includes assets used in the manufacture and assembly of airborne vehicles and their component parts including hydraulic, pneumatic, electrical, and mechanical systems. Does not include assets used in the production of electronic airborne detection, guidance, control, radiation, computation, test, navigation, and communication equipment or the components thereof.	10	7	10
37.31	**Ship and Boat Building Machinery and Equipment:** Includes assets used in the manufacture and repair of ships, boats, caissons, marine drilling rigs, and special fabrications not included in asset classes 37.32 and 37.33. Specifically includes all manufacturing and repairing machinery and equipment, including machinery and equipment used in the operation of assets included in asset class 37.32. Excludes buildings and their structural components.	12	7	12
37.32	**Ship and Boat Building Dry Docks and Land Improvements:** Includes assets used in the manufacture and repair of ships, boats, caissons, marine drilling rigs, and special fabrications not included in asset classes 37.31 and 37.33. Specifically includes floating and fixed dry docks, ship basins, graving docks, shipways, piers, and all other land improvements such as water, sewer, and electric systems. Excludes buildings and their structural components.	16	10	16
37.33	**Ship and Boat Building—Special Tools:** Includes assets defined as special tools such as dies, jigs, molds, patterns, fixtures, gauges, and drawings concerning such special tools used in the activities defined in classes 37.31 and 37.32. Special tools are specifically designed for the production or processing of particular machine components, products, or parts, and have no significant utilitarian value and cannot be adapted to further or different use after changes or improvements are made in the model design of the particular part produced by the special tools. Does not include general purpose small tools such as wrenches and drills, both hand and power-driven, and other general purpose equipment such as conveyors, transfer equipment, and materials handling devices.	6.5	5	6.5
37.41	**Manufacture of Locomotives:** Includes assets used in building or rebuilding railroad locomotives (including mining and industrial locomotives). Does not include assets of railroad transportation companies or assets of companies which manufacture components of locomotives but do not manufacture finished locomotives.	11.5	7	11.5
37.42	**Manufacture of Railroad Cars:** Includes assets used in building or rebuilding railroad freight or passenger cars (including rail transit cars). Does not include assets of railroad transportation companies or assets of companies which manufacture components of railroad cars but do not manufacture finished railroad cars.	12	7	12
39.0	**Manufacture of Athletic, Jewelry, and Other Goods:** Includes assets used in the production of jewelry; musical instruments; toys and sporting goods; motion picture and television films and tapes; and pens, pencils, office and art supplies, brooms, brushes, caskets, etc. **Railroad Transportation:** Classes with the prefix 40 include the assets identified below that are used in the commercial and contract carrying of passengers and freight by rail. Assets of electrified railroads will be classified in a manner corresponding to that set forth below for railroads not independently operated as electric lines. Excludes the assets included in classes with the prefix beginning 00.1 and 00.2 above, and also excludes any non-depreciable assets included in Interstate Commerce Commission accounts enumerated for this class.	12	7	12

Figure 5.2 continued

Credit: IRS "Revenue Procedure 87-56"

Table B-2. **Table of Class Lives and Recovery Periods**

Asset class	Description of assets included	Class Life (in years)	Recovery Periods (in years) GDS (MACRS)	ADS
40.1	**Railroad Machinery and Equipment:** Includes assets classified in the following Interstate Commerce Commission accounts: **Roadway accounts:** (16) Station and office buildings (freight handling machinery and equipment only) (25) TOFC/COFC terminals (freight handling machinery and equipment only) (26) Communication systems (27) Signals and interlockers (37) Roadway machines (44) Shop machinery **Equipment accounts:** (52) Locomotives (53) Freight train cars (54) Passenger train cars (57) Work equipment	14	7	14
40.2	**Railroad Structures and Similar Improvements:** Includes assets classified in the following Interstate Commerce Commission road accounts: (6) Bridges, trestles, and culverts (7) Elevated structures (13) Fences, snowsheds, and signs (16) Station and office buildings (stations and other operating structures only) (17) Roadway buildings (18) Water stations (19) Fuel stations (20) Shops and enginehouses (25) TOFC/COFC terminals (operating structures only) (31) Power transmission systems (35) Miscellaneous structures (39) Public improvements construction	30	20	30
40.3	**Railroad Wharves and Docks:** Includes assets classified in the following Interstate Commerce accounts: (23) Wharves and docks (24) Coal and ore wharves	20	15	20
40.4	**Railroad Track**	10	7	10
40.51	**Railroad Hydraulic Electric Generating Equipment**	50	20	50
40.52	**Railroad Nuclear Electric Generating Equipment**	20	15	20
40.53	**Railroad Steam Electric Generating Equipment**	28	20	28
40.54	**Railroad Steam, Compressed Air, and Other Power Plan Equipment**	28	20	28
41.0	**Motor Transport—Passengers:** Includes assets used in the urban and interurban commercial and contract carrying of passengers by road, except the transportation assets included in classes with the prefix 00.2.	8	5	8
42.0	**Motor Transport—Freight:** Includes assets used in the commercial and contract carrying of freight by road, except the transportation assets included in classes with the prefix 00.2.	8	5	8
44.0	**Water Transportation:** Includes assets used in the commercial and contract carrying of freight and passengers by water except the transportation assets included in classes with the prefix 00.2. Includes all related land improvements.	20	15	20
45.0	**Air Transport:** Includes assets (except helicopters) used in commercial and contract carrying of passengers and freight by air. For purposes of section 1.167(a)-11(d)(2)(iv)(a) of the regulations, expenditures for "repair, maintenance, rehabilitation, or improvement," shall consist of direct maintenance expenses (irrespective of airworthiness provisions or charges) as defined by Civil Aeronautics Board uniform accounts 5200, maintenance burden (exclusive of expenses pertaining to maintenance buildings and improvements) as defined by Civil Aeronautics Board accounts 5300, and expenditures which are not "excluded additions" as defined in section 1.167(a)-11(d)(2)(vi) of the regulations and which would be charged to property and equipment accounts in the Civil Aeronautics Board uniform system of accounts.	12	7	12
45.1	**Air Transport (restricted):** Includes each asset described in the description of class 45.0 which was held by the taxpayer on April 15, 1976, or is acquired by the taxpayer pursuant to a contract which was, on April 15, 1976, and at all times thereafter, binding on the taxpayer. This criterion of classification based on binding contract concept is to be applied in the same manner as under the general rules expressed in section 49(b)(1), (4), (5) and (8) of the Code (as in effect prior to its repeal by the Revenue Act of 1978, section 312(c)(1), (d), 1978-3 C.B. 1, 60).	6	5	6
46.0	**Pipeline Transportation:** Includes assets used in the private, commercial, and contract carrying of petroleum, gas and other products by means of pipes and conveyors. The trunk lines and related storage facilities of integrated petroleum and natural gas producers are included in this class. Excludes initial clearing and grading land improvements as specified in Rev. Rul. 72-403, 1972-2; C.B. 102, but includes all other related land improvements.	22	15	22

Figure 5.2 continued

Credit: IRS "Revenue Procedure 87-56"

Table B-2. **Table of Class Lives and Recovery Periods**

Asset class	Description of assets included	Recovery Periods (in years) Class Life (in years)	GDS (MACRS)	ADS
48.11	**Telephone Communications:** Includes the assets classified below and that are used in the provision of commercial and contract telephonic services such as: **Telephone Central Office Buildings:** Includes assets intended to house central office equipment, as defined in Federal Communications Commission Part 31 Account No. 212 whether section 1245 or section 1250 property.	45	20	45
48.12	**Telephone Central Office Equipment:** Includes central office switching and related equipment as defined in Federal Communications Commission Part 31 Account No. 221. Does not include computer-based telephone central office switching equipment included in class 48.121. Does not include private branch exchange (PBX) equipment.	18	10	18
48.121	**Computer-based Telephone Central Office Switching Equipment:** Includes equipment whose functions are those of a computer or peripheral equipment (as defined in section 168(i)(2)(B) of the Code) used in its capacity as telephone central office equipment. Does not include private exchange (PBX) equipment.	9.5	5	9.5
48.13	**Telephone Station Equipment:** Includes such station apparatus and connections as teletypewriters, telephones, booths, private exchanges, and comparable equipment as defined in Federal Communications Commission Part 31 Account Nos. 231, 232, and 234.	10	7*	10*
48.14	**Telephone Distribution Plant:** Includes such assets as pole lines, cable, aerial wire, underground conduits, and comparable equipment, and related land improvements as defined in Federal Communications Commission Part 31 Account Nos. 241, 242.1, 242.2, 242.3, 242.4, 243, and 244.	24	15	24
48.2	**Radio and Television Broadcastings:** Includes assets used in radio and television broadcasting, except transmitting towers. **Telegraph, Ocean Cable, and Satellite Communications (TOCSC)** includes communications-related assets used to provide domestic and international radio-telegraph, wire-telegraph, ocean-cable, and satellite communications services; also includes related land improvements. If property described in Classes 48.31–48.45 is comparable to telephone distribution plant described in Class 48.14 and used for 2-way exchange of voice and data communication which is the equivalent of telephone communication, such property is assigned a class life of 24 years under this revenue procedure. Comparable equipment does not include cable television equipment used primarily for 1-way communication.	6	5	6
48.31	**TOCSC—Electric Power Generating and Distribution Systems:** Includes assets used in the provision of electric power by generation, modulation, rectification, channelization, control, and distribution. Does not include these assets when they are installed on customers premises.	19	10	19
48.32	**TOCSC—High Frequency Radio and Microwave Systems:** Includes assets such as transmitters and receivers, antenna supporting structures, antennas, transmission lines from equipment to antenna, transmitter cooling systems, and control and amplification equipment. Does not include cable and long-line systems.	13	7	13
48.33	**TOCSC—Cable and Long-line Systems:** Includes assets such as transmission lines, pole lines, ocean cables, buried cable and conduit, repeaters, repeater stations, and other related assets. Does not include high frequency radio or microwave systems.	26.5	20	26.5
48.34	**TOCSC—Central Office Control Equipment:** Includes assets for general control, switching, and monitoring of communications signals including electromechanical switching and channeling apparatus, multiplexing equipment patching and monitoring facilities, in-house cabling, teleprinter equipment, and associated site improvements.	16.5	10	16.5
48.35	**TOCSC—Computerized Switching, Channeling, and Associated Control Equipment:** Includes central office switching computers, interfacing computers, other associated specialized control equipment, and site improvements.	10.5	7	10.5
48.36	**TOCSC—Satellite Ground Segment Property:** Includes assets such as fixed earth station equipment, antennas, satellite communications equipment, and interface equipment used in satellite communications. Does not include general purpose equipment or equipment used in satellite space segment property.	10	7	10
48.37	**TOCSC—Satellite Space Segment Property:** Includes satellites and equipment used for telemetry, tracking, control, and monitoring when used in satellite communications.	8	5	8
48.38	**TOCSC—Equipment Installed on Customer's Premises:** Includes assets installed on customer's premises, such as computers, terminal equipment, power generation and distribution systems, private switching center, teleprinters, facsimile equipment and other associated and related equipment.	10	7	10
48.39	**TOCSC—Support and Service Equipment:** Includes assets used to support but not engage in communications. Includes store, warehouse and shop tools, and test and laboratory assets. **Cable Television (CATV):** Includes communications-related assets used to provide cable television community antenna television services. Does not include assets used to provide subscribers with two-way communications services.	13.5	7	13.5

* Property described in asset guideline class 48.13 which is qualified technological equipment as defined in section 168(i)(2) is assigned a 5-year recovery period.

Figure 5.2 continued

Credit: IRS "Revenue Procedure 87-56"

Table B-2. **Table of Class Lives and Recovery Periods**

Asset class	Description of assets included	Recovery Periods (in years) Class Life (in years)	GDS (MACRS)	ADS
48.41	**CATV—Headend:** Includes assets such as towers, antennas, preamplifiers, converters, modulation equipment, and program non-duplication systems. Does not include headend buildings and program origination assets.	11	7	11
48.42	**CATV—Subscriber Connection and Distribution Systems:** Includes assets such as trunk and feeder cable, connecting hardware, amplifiers, power equipment, passive devices, directional taps, pedestals, pressure taps, drop cables, matching transformers, multiple set connector equipment, and convertors.	10	7	10
48.43	**CATV—Program Origination:** Includes assets such as cameras, film chains, video tape recorders, lighting, and remote location equipment excluding vehicles. Does not include buildings and their structural components.	9	5	9
48.44	**CATV—Service and Test:** Includes assets such as oscilloscopes, field strength meters, spectrum analyzers, and cable testing equipment, but does not include vehicles.	8.5	5	8.5
48.45	**CATV—Microwaave Systems:** Inlcudes assets such as towers, antennas, transmitting and receiving equipment, and broad band microwave assets is used in the provision of cable television services. Does not include assets used in the provision of common carrier services.	9.5	5	9.5
49.11	**Electric, Gas, Water and Steam, Utility Services:** Includes assets used in the production, transmission and distribution of electricity, gas, steam, or water for sale including related land improvements. **Electric Utility Hydraulic Production Plant:** Includes assets used in the hydraulic power production of electricity for sale, including related land improvements, such as dams, flumes, canals, and waterways.	50	20	50
49.12	**Electric Utility Nuclear Production Plant:** Includes assets used in the nuclear power production and electricity for sale and related land improvements. Does not include nuclear fuel assemblies.	20	15	20
49.121	**Electric Utility Nuclear Fuel Assemblies:** Includes initial core and replacement core nuclear fuel assemblies (i.e., the composite of fabricated nuclear fuel and container) when used in a boiling water, pressurized water, or high temperature gas reactor used in the production of electricity. Does not include nuclear fuel assemblies used in breader reactors.	5	5	5
49.13	**Electric Utility Steam Production Plant:** Includes assets used in the steam power production of electricity for sale, combusion turbines operated in a combined cycle with a conventional steam unit and related land improvements. Also includes package boilers, electric generators and related assets such as electricity and steam distribution systems as used by a waste reduction and resource recovery plant if the steam or electricity is normally for sale to others.	28	20	28
49.14	**Electric Utility Transmission and Distribution Plant:** Includes assets used in the transmission and distribution of electricity for sale and related land improvements. Excludes initial clearing and grading land improvements as specified in Rev. Rul. 72-403, 1972-2 C.B. 102.	30	20	30
49.15	**Electric Utility Combustion Turbine Production Plant:** Includes assets used in the production of electricity for sale by the use of such prime movers as jet engines, combustion turbines, diesel engines, gasoline engines, and other internal combustion engines, their associated power turbines and/or generators, and related land improvements. Does not include combustion turbines operated in a combined cycle with a conventional steam unit.	20	15	20
49.21	**Gas Utility Distribution Facilities:** Includes gas water heaters and gas conversion equipment installed by utility on customers' premises on a rental basis.	35	20	35
49.221	**Gas Utility Manufactured Gas Production Plants:** Includes assets used in the manufacture of gas having chemical and/or physical properties which do not permit complete interchangeability with domestic natural gas. Does not include gas-producing systems and related systems used in waste reduction and resource recovery plants which are elsewhere classified.	30	20	30
49.222	**Gas Utility Substitute Natural Gas (SNG) Production Plant (naphtha or lighter hydrocarbon feedstocks):** Includes assets used in the catalytic conversion of feedstocks or naphtha or lighter hydrocarbons to a gaseous fuel which is completely interchangeable with domestic natural gas.	14	7	14
49.223	**Substitute Natural Gas—Coal Gasification:** Includes assets used in the manufacture and production of pipeline quality gas from coal using the basic Lurgi process with advanced methanation. Includes all process plant equipment and structures used in this coal gasification process and all utility assets such as cooling systems, water supply and treatment facilities, and assets used in the production and distribution of electricity and steam for use by the taxpayer in a gasification plant and attendant coal mining site processes but not for assets used in the production and distribution of electricity and steam for sale to others. Also includes all other related land improvements. Does not include assets used in the direct mining and treatment of coal prior to the gasification process itself.	18	10	18
49.23	**Natural Gas Production Plant**	14	7	14
49.24	**Gas Utility Trunk Pipelines and Related Storage Facilities:** Excluding initial clearing and grading land improvements as specified in Rev. Rul. 72-40.	22	15	22
49.25	**Liquefied Natural Gas Plant:** Includes assets used in the liquefaction, storage, and regasification of natural gas including loading and unloading connections, instrumentation equipment and controls, pumps, vaporizers and odorizers, tanks, and related land improvements. Also includes pipeline interconnections with gas transmission lines and distribution systems and marine terminal facilities.	22	15	22

Figure 5.2 continued

Credit: IRS "Revenue Procedure 87-56"

Table B-2. **Table of Class Lives and Recovery Periods**

Asset class	Description of assets included	Recovery Periods (in years) Class Life (in years)	GDS (MACRS)	ADS
49.3	**Water Utilities:** Includes assets used in the gathering, treatment, and commercial distribution of water.	50	20***	50
49.4	**Central Steam Utility Production and Distribution:** Includes assets used in the production and distribution of steam for sale. Does not include assets used in waste reduction and resource recovery plants which are elsewhere classified.	28	20	28
49.5	**Waste Reduction and Resource Recovery Plants:** Includes assets used in the conversion of refuse or other solid waste or biomass to heat or to a solid, liquid, or gaseous fuel. Also includes all process plant equipment and structures at the site used to receive, handle, collect, and process refuse or other solid waste or biomass in a waterwall, combustion system, oil or gas pyrolysis system, or refuse derived fuel system to create hot water, gas, steam and electricity. Includes material recovery and support assets used in refuse or solid refuse or solid waste receiving, collecting, handling, sorting, shredding, classifying, and separation systems. Does not include any package boilers, or electric generators and related assets such as electricity, hot water, steam and manufactured gas production plants classified in classes 00.4, 49.13, 49.221, and 49.4. Does include, however, all other utilities such as water supply and treatment facilities, ash handling and other related land improvements of a waste reduction and resource recovery plant.	10	7	10
50.	**Municipal Wastewater Treatment Plant**	24	15	24
51.	**Municipal Sewer**	50	20***	50
57.0	**Distributive Trades and Services:** Includes assets used in wholesale and retail trade, and personal and professional services. Includes section 1245 assets used in marketing petroleum and petroleum products.	9	5	9*
57.1	**Distributive Trades and Services—Billboard, Service Station Buildings and Petroleum Marketing Land Improvements:** Includes section 1250 assets, including service station buildings and depreciable land improvements, whether section 1245 property or section 1250 property, used in the marketing of petroleum and petroleum products, but not including any of these facilities related to petroleum and natural gas trunk pipelines. Includes car wash buildings and related land improvements. Includes billboards, whether such assets are section 1245 property or section 1250 property. Excludes all other land improvements, buildings and structural components as defined in section 1.48-1(e) of the regulations. See *Gas station convenience stores* in chapter 3.	20	15	20
79.0	**Recreation:** Includes assets used in the provision of entertainment services on payment of a fee or admission charge, as in the operation of bowling alleys, billiard and pool establishments, theaters, concert halls, and miniature golf courses. Does not include amusement and theme parks and assets which consist primarily of specialized land improvements or structures, such as golf courses, sports stadia, race tracks, ski slopes, and buildings which house the assets used in entertainment services.	10	7	10
80.0	**Theme and Amusement Parks:** Includes assets used in the provision of rides, attractions, and amusements in activities defined as theme and amusement parks, and includes appurtenances associated with a ride, attraction, amusement or theme setting within the park such as ticket booths, facades, shop interiors, and props, special purpose structures, and buildings other than warehouses, administration buildings, hotels, and motels. Includes all land improvements for or in support of park activities (e.g., parking lots, sidewalks, waterways, bridges, fences, landscaping, etc.), and support functions (e.g., food and beverage retailing, souvenir vending and other nonlodging accommodations) if owned by the park and provided exclusively for the benefit of park patrons. Theme and amusement parks are defined as combinations of amusements, rides, and attractions which are permanently situated on park land and open to the public for the price of admission. This guideline class is a composite of all assets used in this industry except transportation equipment (general purpose trucks, cars, airplanes, etc., which are included in asset guideline classes with the prefix 00.2), assets used in the provision of administrative services (asset classes with the prefix 00.1) and warehouses, administration buildings, hotels and motels.	12.5	7	12.5
	Certain Property for Which Recovery Periods Assigned			
	A. Personal Property With No Class Life		7	12
	Section 1245 Real Property With No Class Life		7	40
	B. Qualified Technological Equipment, as defined in section 168(i)(2).	**	5	5
	C. Property Used in Connection with Research and Experimentation referred to in section 168(e)(3)(B).	**	5	class life if no class life—12
	D. Alternative Energy Property described in sections 48(1)(3)(viii) or (iv), or section 48(1)(4) of the Code.	**	5	class life if no class life—12
	E. Biomass property described in section 48(1)(15) and is a qualifying small production facility within the meaning of section 3(17)(c) of the Federal Power Act (16 U.S.C. 796(17)(C)), as in effect on September 1, 1986.	**	5	class life if no class life—12

* Any high technology medical equipment as defined in section 168(i)(2)(C) which is described in asset guideline class 57.0 is assigned a 5-year recovery period for the alternate MACRS method.

** The class life (if any) of property described in classes B, C, D, or E is determined by reference to the asset guideline classes. If an item of property described in paragraphs B, C, D, or E is not described in any asset guideline class, such item of property has no class life.

*** Use straight line over 25 years if placed in service after June 12, 1996, unless placed in service under a binding contract in effect before June 10, 1996, and at all times until placed in service.

Figure 5.2 continued

Credit: IRS "Revenue Procedure 87-56"

ASSET CLASSIFICATION TABLE

ASSET	CASE/RULING	CITE	CIRCUIT	BUSINESS CLASSIFICATION FOR DEPRECIATION PURPOSES	CLASSIFICATION	NOTES
ACOUSTICAL CEILINGS	HOSPITAL CORP OF AMERICA	109 TC 21	6TH	HOSPITAL	STRUCTURAL COMPONENT	
AIR CONDITIONING-INDIVIDUAL UNITS	LETTER RULING 7102296400A	LET RUL 7102296400A, FEBRUARY 26, 1971	N/A	MANUFACTURER/WAREHOUSING	OTHER PROPERTY	IF SOLE JUSTIFICATION TEST MET
AIR HANDLING EQUIPMENT	LETTER RULING 7102296400A	LET RUL 7102296400A, FEBRUARY 26, 1971	N/A	MANUFACTURER/WAREHOUSING		
AIR HANDLING EQUIPMENT	REVENUE RULING 69-558	REV RUL 69-558, 1969-2 CB 4	N/A	MANUFACTURER-FABRICS/YARNS	OTHER PROPERTY	SOLE JUSTIFICATION TEST MET
AIR MAKE-UP UNITS-AIR HANDLING EQUIPMENT	LETTER RULING 7102296400A	LET RUL 7102296400A, FEBRUARY 26, 1971	N/A	MANUFACTURER/WAREHOUSING	OTHER PROPERTY	LOCALIZED IN A FACTORY OR RELATED TO SPECIFIC MACHINERY OR PROCESSES
ALLIANCE WALL PANELS	LA PETITE ACADEMY	95-1 USTC 50,193	8 TH	CHILD CARE CENTERS	STRUCTURAL COMPONENT	
ASPHALT ROADWAYS	REVENUE RULING 80-93	REV RUL 80-93, 1980-1 CB 50	N/A	LESSORS-MOBILE HOME PARK	INHERENTLY PERMANENT STRUCTURE	
ASSOCIATED WIRING	ILLINOIS CEREAL MILLS	86-1 USTC ¶9371	7TH	CEREAL MANUFACTURER	OTHER PROPERTY	RELATES TO MANUFACTURING EQUIPMENT
BACKFILLED LAKE	REVENUE RULING 77-270	REV RUL 77-270, 1977-2 CB 79	N/A	STEEL PRODUCER	LAND	ASSOCIATED WITH THE CREATION OF LAND
BANK VAULT	REVENUE RULING 65-79			BANK	TANGIBLE PERSONAL PROPERTY	
BATHROOM ACCESSORIES	HOSPITAL CORP OF AMERICA	109 TC 21	6TH	HOSPITAL	STRUCTURAL COMPONENT	
BATHROOM MIRROR	LA PETITE ACADEMY	95-1 USTC 50,193	8 TH	CHILD CARE CENTERS	STRUCTURAL COMPONENT	
BATHTUBS	TREAS. REG. 1.48-1	TREAS. REG. 1.48-1	N/A	ALL BUSINESS	STRUCTURAL COMPONENT	
BERMS	REVENUE RULING 77-270	REV RUL 77-270, 1977-2 CB 79	N/A	STEEL PRODUCER	LAND	ASSOCIATED WITH THE CREATION OF LAND
BEVERAGE BARS	SENATE RPT 95-1263	SEN RPT 95-1263, 1978-3 CB 315	N/A	ALL BUSINESS	TANGIBLE PERSONAL PROPERTY	
BILLBOARDS/SIGNS	ALABAMA DISPLAYS			ADVERTISING		
BILLBOARDS/SIGNS	NATIONAL ADVERTISING			ADVERTISING		
BILLBOARDS/SIGNS	WHITECO INDUSTRIES			ADVERTISING	TANGIBLE PERSONAL PROPERTY	BASED ON CONSTRUCTION METHOD
BLASTING	EASTWOOD MALL	95-1 USTC 50,236	6TH	MALL OPERATOR	LAND	CREATION OF USABLE LAND
BLINDS	LETTER RULING 8848039	LET RUL 8848039 SEPTEMBER 12, 1988	N/A	LESSORS OF REAL PROPERTY	TANGIBLE PERSONAL PROPERTY	
BLOWERS-AIR HANDLING EQUIPMENT	LETTER RULING 7102296400A	LET RUL 7102296400A, FEBRUARY 26, 1971	N/A	MANUFACTURER/WAREHOUSING	OTHER PROPERTY	LOCALIZED IN A FACTORY OR RELATED TO SPECIFIC MACHINERY OR PROCESSES
BOILERS	HOSPITAL CORP OF AMERICA	109 TC 21	6TH	HOSPITAL	STRUCTURAL COMPONENT	
BOOTHS FOR SEATING	SENATE RPT 95-1263	SEN RPT 95-1263, 1978-3 CB 315	N/A	ALL BUSINESS	TANGIBLE PERSONAL PROPERTY	
BRIDGES	REVENUE PROCEDURE 87-56	REV PROC 87-56, 1987-2 CB 674	N/A	ALL BUSINESS	INHERENTLY PERMANENT STRUCTURE	
BYPASS DOORS	LA PETITE ACADEMY	95-1 USTC 50,193	8 TH	CHILD CARE CENTERS	STRUCTURAL COMPONENT	

Figure 5.3

ASSET CLASSIFICATION TABLE

ASSET	CASE/RULING	CITE	CIRCUIT	BUSINESS CLASSIFICATION FOR DEPRECIATION PURPOSES	CLASSIFICATION	NOTES
CABINET HARDWARE	METRO NATIONAL CORPORATION	TC MEMO 1987-38	5TH	LESSORS OF REAL PROPERTY	TANGIBLE PERSONAL PROPERTY	
CABINETS-FREE STANDING	METRO NATIONAL CORPORATION	TC MEMO 1987-38	5TH	LESSORS OF REAL PROPERTY	TANGIBLE PERSONAL PROPERTY	
CABINETS-MOUNTED	METRO NATIONAL CORPORATION	TC MEMO 1987-38	5TH	LESSORS OF REAL PROPERTY	TANGIBLE PERSONAL PROPERTY	
CABINETS-WITH SINKS	METRO NATIONAL CORPORATION	TC MEMO 1987-38	5TH	LESSORS OF REAL PROPERTY	TANGIBLE PERSONAL PROPERTY	
CANALS	REVENUE PROCEDURE 87-56	REV PROC 87-56, 1987-2 CB 674	N/A	ALL BUSINESS	INHERENTLY PERMANENT STRUCTURE	
CANOPIES	LETTER RULING 7102296400A	LET RUL 7102296400A, FEBRUARY 26, 1971	N/A	MANUFACTURER/WAREHOUSING	INHERENTLY PERMANENT STRUCTURE	
CARPETING	SENATE RPT 95-1263	SEN RPT 95-1263, 1978-3 CB 315	N/A	ALL BUSINESS	TANGIBLE PERSONAL PROPERTY	
CARPETING-REMOVABLE	HOSPITAL CORP OF AMERICA	109 TC 21	6TH	HOSPITAL	TANGIBLE PERSONAL PROPERTY	
CARPETING-WALL TO WALL	REVENUE RULING 67-349	REV RUL 67-349 1967-2 CB 48	N/A	MOTEL OPERATOR	TANGIBLE PERSONAL PROPERTY	
CARWASH EQUIPMENT	SCHRUM	94-2 USTC 50451	4TH			
CATWALKS	REVENUE RULING 74-391	REV RUL 74-391, 1974-2 CB 9	N/A	MANUFACTURING	TANGIBLE PERSONAL PROPERTY	INSTALLED TO SERVICE EQUIPMENT
CATWALKS	REVENUE RULING 74-391	REV RUL 74-391, 1974-2 CB 9	N/A	MANUFACTURING	STRUCTURAL COMPONENT	INSTALLED TO SERVICE BUILDING FUNCTIONS
CATWALKS	LETTER RULING 7102296400A	LET RUL 7102296400A, FEBRUARY 26, 1971	N/A	MANUFACTURER/WAREHOUSING	TANGIBLE PERSONAL PROPERTY	RELATES TO ITEMS OF MACHINERY
CATWALKS/STAIRWAYS	LETTER RULING 7102296400A	LET RUL 7102296400A, FEBRUARY 26, 1971	N/A	MANUFACTURER/WAREHOUSING	STRUCTURAL COMPONENT	SUBJECT TO EMPLOYEES USAGE GENERALLY
CEILING GRIDS	METRO NATIONAL CORPORATION	TC MEMO 1987-38	5TH	LESSORS OF REAL PROPERTY	STRUCTURAL COMPONENT	
CEILINGS	TREAS. REG. 1.48-1	TREAS. REG. 1.48-1	N/A	ALL BUSINESS	STRUCTURAL COMPONENT	
CEILINGS-FALSE	FSA 200203009	FSA 200203009	N/A	CASINO	STRUCTURAL COMPONENT	
CHANDELIER	LETTER RULING 8002015	LETTER RULING 8002015, OCTOBER 1, 1979	N/A	RESTAURANT	STRUCTURAL COMPONENT	PROVIDED GENERAL BUILDING LIGHTING
CHANDELIER	SHONEYS SOUTH	TC MEMO 1984-413	6TH	RESTAURANT	TANGIBLE PERSONAL PROPERTY	
CHILLERS-AIR HANDLING	REVENUE RULING 69-558	REV RUL 69-558, 1969-2 CB 4	N/A	YARNS-FABRICS/YARNS	OTHER PROPERTY	SOLE JUSTIFICATION TEST MET
CHIMNEYS	TREAS. REG. 1.48-1	TREAS. REG. 1.48-1	N/A	ALL BUSINESS	STRUCTURAL COMPONENT	
CIRCUIT BREAKERS	ILLINOIS CEREAL MILLS	86-1 USTC ¶9371	7TH	CEREAL MANUFACTURER	OTHER PROPERTY	RELATES TO MANUFACTURING EQUIPMENT
CLEAR & GRUB	EASTWOOD MALL	95-1 USTC 50,236	6TH	MALL OPERATOR	LAND	CREATION OF USABLE LAND

Figure 5.3 continued

ASSET CLASSIFICATION TABLE

ASSET	CASE/RULING	CITE	CIRCUIT	BUSINESS CLASSIFICATION FOR DEPRECIATION PURPOSES	CLASSIFICATION	NOTES
CLEARING-SITE WORK	REVENUE RULING 80-93	REV RUL 80-93, 1980-1 CB 50	N/A	LESSORS-MOBILE HOME PARK	LAND	PREPARES LAND FOR ITS INTENDED USE
COMMUNICATION EQUIPMENT	REVENUE PROCEDURE 87-56	REV PROC 87-56, 1987-2 CB 674	N/A	ALL BUSINESS	TANGIBLE PERSONAL PROPERTY	
COMPUTERS-GENERAL BUSINESS	REVENUE PROCEDURE 87-56	REV PROC 87-56, 1987-2 CB 674	N/A	ALL BUSINESS	TANGIBLE PERSONAL PROPERTY	
CONCRETE SIDEWALKS	REVENUE RULING 80-93	REV RUL 80-93, 1980-1 CB 50	N/A	LESSORS-MOBILE HOME PARK	INHERENTLY PERMANENT STRUCTURE	
CONTROL SYSTEMS-AIR HANDLING	REVENUE RULING 69-558	REV RUL 69-558, 1969-2 CB 4	N/A	YARNS-FABRICS/YARNS	OTHER PROPERTY	SOLE JUSTIFICATION TEST MET
COPPER BUS DUCT-ELECTRICAL	AC MONK & COMPANY	82-2 USTC ¶9551	4TH	TOBACCO PROCESSING	BOTH	COULD BE REASONABLY ADAPTED TO OTHER USES- REMANDED TO LOWER COURT FOR ITEM BY ITEM DETERMINATION
COPY MACHINES-NETWORKED	REVENUE PROCEDURE 87-56	REV PROC 87-56, 1987-2 CB 674	N/A	ALL BUSINESS	TANGIBLE PERSONAL PROPERTY	
CUT & FILL	EASTWOOD MALL	95-1 USTC 50,236	6TH	MALL OPERATOR	LAND	CREATION OF USABLE LAND
CUTTING-SITE WORK	REVENUE RULING 80-93	REV RUL 80-93, 1980-1 CB 50	N/A	LESSORS-MOBILE HOME PARK	LAND	PREPARES LAND FOR ITS INTENDED USE
DECORATIVE LIGHTING	FSA 200203009	FSA 200203009	N/A	CASINO	TANGIBLE PERSONAL PROPERTY	
DECORATIVE LIGHTING	LETTER RULING 8002015	LETTER RULING 8002015, OCTOBER 1, 1979	N/A	RESTAURANT	STRUCTURAL COMPONENT	PROVIDED GENERAL BUILDING LIGHTING
DEMOLITION	REVENUE PROCEDURE 95-27	REV PROC 95-27, 1995-1 CB 704	N/A	ALL BUSINESS	LAND	LESS THAN 75% OF STRUCTURE REMAINED
DEMOLITION	REVENUE PROCEDURE 95-27	REV PROC 95-27, 1995-1 CB 704	N/A	ALL BUSINESS	ALLOCABLE EXPENDITURE	75% OR MORE OF THE STRUCTURE REMAINS
DEMOLITION ANY STRUCTURE	IRC SECTION 280B	IRC SECTION 280B(a)	N/A	ALL BUSINESS	LAND	
DEMOLITION-BUILDINGS	TREAS. REG. 1.165-3		N/A	ALL BUSINESS	LAND	
DESKS	REVENUE PROCEDURE 87-56	REV PROC 87-56, 1987-2 CB 674	N/A	ALL BUSINESS	TANGIBLE PERSONAL PROPERTY	
DOCKS	REVENUE PROCEDURE 87-56	REV PROC 87-56, 1987-2 CB 674	N/A	ALL BUSINESS	INHERENTLY PERMANENT STRUCTURE	
DOCKS	LETTER RULING 7102296400A	LET RUL 7102296400A, FEBRUARY 26, 1971	N/A	MANUFACTURER/WAREHOUSING	STRUCTURAL COMPONENT	ENCLOSED WITH THE BUILDING STRUCTURE
DOCKS	LETTER RULING 7102296400A	LET RUL 7102296400A, FEBRUARY 26, 1971	N/A	MANUFACTURER/WAREHOUSING	OTHER PROPERTY	EXTENDS BEYOND THE BUILDING STRUCTURE
DOOR LOCKS-COMPUTERIZED	FSA 200203009	FSA 200203009	N/A	CASINO	STRUCTURAL COMPONENT	
DOORS	LA PETITE ACADEMY	95-1 USTC 50,193	8 TH	CHILD CARE CENTERS	STRUCTURAL COMPONENT	
DOORS	TREAS. REG. 1.48-1	TREAS. REG. 1.48-1	N/A	ALL BUSINESS	STRUCTURAL COMPONENT	
DOORS-OVERHEAD	CONSOLIDATED FREIGHTWAYS	83-1 USTC 9420		TRANSPORTATION/WAREHOUSE	STRUCTURAL COMPONENT	
DRAINAGE FACILITIES	REVENUE PROCEDURE 87-56	REV PROC 87-56, 1987-2 CB 674	N/A	ALL BUSINESS	INHERENTLY PERMANENT STRUCTURE	

Figure 5.3 continued

ASSET CLASSIFICATION TABLE

ASSET	CASE/RULING	CITE	CIRCUIT	BUSINESS CLASSIFICATION FOR DEPRECIATION PURPOSES	CLASSIFICATION	NOTES
DRIVE UP BOOTH	REVENUE RULING 65-79			BANK	STRUCTURAL COMPONENT	RESEMBLED A BUILDING
DRIVE UP WINDOW	REVENUE RULING 65-79			BANK	TANGIBLE PERSONAL PROPERTY	
DRIVEWAYS	LETTER RULING 8848039	LET RUL 8848039 SEPTEMBER 12, 1988	N/A	LESSORS OF REAL PROPERTY	INHERENTLY PERMANENT STRUCTURE	
DUCTWORK-AIR HANDLING	REVENUE RULING 69-558	REV RUL 69-558, 1969-2 CB 4	N/A	YARNS-FABRICS/YARNS	OTHER PROPERTY	SOLE JUSTIFICATION TEST MET
DUMPSTER ENCLOSURE	LA PETITE ACADEMY	95-1 USTC 50,193	8 TH	CHILD CARE CENTERS	INHERENTLY PERMANENT STRUCTURE	
DUPLICATING EQUIPMENT	REVENUE PROCEDURE 87-56	REV PROC 87-56, 1987-2 CB 674	N/A	ALL BUSINESS	TANGIBLE PERSONAL PROPERTY	
DUST COLLECTORS-AIR HANDLING EQUIPMENT	LETTER RULING 7102296400A	LET RUL 7102296400A, FEBRUARY 26, 1971	N/A	MANUFACTURER/WAREHOUSING	OTHER PROPERTY	LOCALIZED IN A FACTORY OR RELATED TO SPECIFIC MACHINERY OR PROCESSES
DUTCH DOORS	LA PETITE ACADEMY	95-1 USTC 50,193	8 TH	CHILD CARE CENTERS	STRUCTURAL COMPONENT	
ELECTRIC OUTLETS	CENTRAL CITRUS	58 TC 37	9 TH	FOOD MANUFACTURER	STRUCTURAL COMPONENT	RELATES TO THE OPERATION OF THE BUILDING
ELECTRIC PANELS-KITCHEN	MORRISON	90-1 USTC 50,034	11 TH	RESTAURANTS	TANGIBLE PERSONAL PROPERTY	FUNCTIONAL ALLOCATION USED
ELECTRIC PRIMARY DISTRIBUTION	MORRISON	90-1 USTC 50,034	11 TH	RESTAURANTS	TANGIBLE PERSONAL PROPERTY	FUNCTIONAL ALLOCATION USED
ELECTRIC TIMERS	CENTRAL CITRUS	58 TC 37	9 TH	FOOD MANUFACTURER	TANGIBLE PERSONAL PROPERTY	RELATES TO EQUIPMENT USED IN MANUFACTURING
ELECTRIC WIRING	TREAS. REG. 1.48-1	TREAS. REG. 1.48-1	N/A	ALL BUSINESS	STRUCTURAL COMPONENT	
ELECTRICAL DISTRIBUTION SYSTEM	CENTRAL CITRUS	58 TC 37	9 TH	FOOD MANUFACTURER	TANGIBLE PERSONAL PROPERTY	RELATES TO EQUIPMENT USED IN MANUFACTURING
ELECTRICAL DISTRIBUTIONS SYSTEM	ILLINOIS CEREAL MILLS	86-1 USTC ¶9371	7TH	CEREAL MANUFACTURER	OTHER PROPERTY	RELATES TO MANUFACTURING EQUIPMENT
ELECTRICAL OUTLETS-GUESTROOMS	FSA 200203009	FSA 200203009	N/A	CASINO	STRUCTURAL COMPONENT	
ELECTRICAL SWITCHGEAR/CABLES	SCOTT PAPER	74 TC 137		MANUFACTURER-PAPER PRODUCTS	STRUCTURAL COMPONENT	FUNCTIONAL ALLOCATION USED
ELECTRICAL SWITCHGEAR/CABLES	SCOTT PAPER	74 TC 137		MANUFACTURER-PAPER PRODUCTS	TANGIBLE PERSONAL PROPERTY	FUNCTIONAL ALLOCATION USED
ELECTRICAL SYSTEM-PLANT	AC MONK & COMPANY	82-2 USTC ¶9551	4TH	TOBACCO PROCESSING	BOTH	COULD BE REASONABLY ADAPTED TO OTHER USES- REMANDED TO LOWER COURT FOR ITEM BY ITEM DETERMINATION
ELECTRICAL SYSTEMS-EQUIPMENT	HOSPITAL CORP OF AMERICA	109 TC 21	6TH	HOSPITAL	TANGIBLE PERSONAL PROPERTY	
ELECTRIC-DEAD FRONT PANEL	CENTRAL CITRUS	58 TC 37	9 TH	FOOD MANUFACTURER	STRUCTURAL COMPONENT	RELATES TO THE OPERATION OF THE BUILDING

Figure 5.3 continued

ASSET CLASSIFICATION TABLE

ASSET	CASE/RULING	CITE	CIRCUIT	BUSINESS CLASSIFICATION FOR DEPRECIATION PURPOSES	CLASSIFICATION	NOTES
ELECTRIC-SPECIAL HOOKUPS	REVENUE RULING 66-299	REV RUL 66-299, 1966-2 CB 14	N/A	MANUFACTURER	TANGIBLE PERSONAL PROPERTY	RELATES TO MANUFACTURING EQUIPMENT
ELECTRIC-TRANSFORMER	CENTRAL CITRUS	58 TC 37	9 TH	FOOD MANUFACTURER	STRUCTURAL COMPONENT	RELATES TO THE OPERATION OF THE BUILDING
ELECTRIC-UNDERGROUND	REVENUE RULING 80-93	REV RUL 80-93, 1980-1 CB 50	N/A	LESSORS-MOBILE HOME PARK	LAND	NOT OWNED BY THE TAX PAYER
ELEVATORS	TREAS. REG. 1.48-1	TREAS. REG. 1.48-1	N/A	ALL BUSINESS	STRUCTURAL COMPONENT	
EMERGENCY GENERATORS	FSA 200203009	FSA 200203009	N/A	CASINO	STRUCTURAL COMPONENT	SUPPORTS BUILDING FUNCTIONS ALLOCATION USED
EMERGENCY GENERATORS	FSA 200203009	FSA 200203009	N/A	CASINO	TANGIBLE PERSONAL PROPERTY	SUPPORTS BUSINESS EQUIPMENT ALLOCATION USED
ESCALATORS	TREAS. REG. 1.48-1	TREAS. REG. 1.48-1	N/A	ALL BUSINESS	STRUCTURAL COMPONENT	
EXCESS COSTS	REVENUE RULING 69-558	REV RUL 69-558, 1969-2 CB 4	N/A	YARNS-FABRICS/YARNS	STRUCTURAL COMPONENT	
EXPLOSION HATCHES-SAFETY EQUIPMENT	LETTER RULING 7102296400A	LET RUL 7102296400A, FEBRUARY 26, 1971	N/A	MANUFACTURER/WAREHOUSING	OTHER PROPERTY	LOCALIZED IN A FACTORY OR RELATED TO SPECIFIC MACHINERY OR PROCESSES
EXTERIOR BUILDING LIGHTING	SENATE RPT 95-1263	SEN RPT 95-1263, 1978-3 CB 315	N/A	ALL BUSINESS	TANGIBLE PERSONAL PROPERTY	
EXTERIOR LIGHTING	LA PETITE ACADEMY	95-1 USTC 50,193	8 TH	CHILD CARE CENTERS	STRUCTURAL COMPONENT	PROVIDED LIGHTING FOR SIDEWALKS
EXTERIOR ORNAMENTATION	SENATE RPT 95-1263	SEN RPT 95-1263, 1978-3 CB 315	N/A	ALL BUSINESS	TANGIBLE PERSONAL PROPERTY	NO MORE THAN INCIDENTAL RELATIONSHIP TO OPERATION AND MAINTENANCE OF BUILDINGS
FALSE BALCONIES	SENATE RPT 95-1263	SEN RPT 95-1263, 1978-3 CB 315	N/A	ALL BUSINESS	TANGIBLE PERSONAL PROPERTY	NO MORE THAN INCIDENTAL RELATIONSHIP TO OPERATION AND MAINTENANCE OF BUILDINGS
FALSE CEILINGS	METRO NATIONAL CORPORATION	TC MEMO 1987-38	5TH	LESSORS OF REAL PROPERTY	STRUCTURAL COMPONENT	
FALSE FLOOR	REVENUE RULING 74-391	REV RUL 74-391, 1974-2 CB 9	N/A	MANUFACTURING	TANGIBLE PERSONAL PROPERTY	SPECIFIC USE FOR COMPUTERS
FENCES	REVENUE PROCEDURE 87-56	REV PROC 87-56, 1987-2 CB 674	N/A	ALL BUSINESS	INHERENTLY PERMANENT STRUCTURE	
FENCING	LETTER RULING 8848039	LET RUL 8848039 SEPTEMBER 12, 1988	N/A	LESSORS OF REAL PROPERTY	INHERENTLY PERMANENT STRUCTURE	
FILES	REVENUE PROCEDURE 87-56	REV PROC 87-56, 1987-2 CB 674	N/A	ALL BUSINESS	TANGIBLE PERSONAL PROPERTY	
FILLED-IN LAKE	REVENUE RULING 77-270	REV RUL 77-270, 1977-2 CB 79	N/A	STEEL PRODUCER	LAND	ASSOCIATED WITH THE CREATION OF LAND
FILLING-SITE WORK	REVENUE RULING 80-93	REV RUL 80-93, 1980-1 CB 50	N/A	LESSORS-MOBILE HOME PARK	LAND	PREPARES LAND FOR ITS INTENDED USE
FIRE CURTAINS-SAFETY EQUIPMENT	LETTER RULING 7102296400A	LET RUL 7102296400A, FEBRUARY 26, 1971	N/A	MANUFACTURER/WAREHOUSING	OTHER PROPERTY	LOCALIZED IN A FACTORY OR RELATED TO SPECIFIC MACHINERY OR PROCESSES
FIRE DETECTION-SAFETY EQUIPMENT	LETTER RULING 7102296400A	LET RUL 7102296400A, FEBRUARY 26, 1971	N/A	MANUFACTURER/WAREHOUSING	OTHER PROPERTY	LOCALIZED IN A FACTORY OR RELATED TO SPECIFIC MACHINERY OR PROCESSES

Figure 5.3 continued

ASSET CLASSIFICATION TABLE

ASSET	CASE/RULING	CITE	CIRCUIT	BUSINESS CLASSIFICATION FOR DEPRECIATION PURPOSES	CLASSIFICATION	NOTES
FIRE ESCAPES	TREAS. REG. 1.48-1	TREAS. REG. 1.48-1	N/A	ALL BUSINESS	STRUCTURAL COMPONENT	
FIRE PROTECTION SYSTEM	LA PETITE ACADEMY	95-1 USTC 50,193	8 TH	CHILD CARE CENTERS	STRUCTURAL COMPONENT	SERVES ENTIRE BUILDING
FLOODLAMPS	CENTRAL CITRUS	58 TC 37	9 TH	FOOD MANUFACTURER	TANGIBLE PERSONAL PROPERTY	RELATES TO EQUIPMENT USED IN MANUFACTURING
FLOOR COVERINGS-REMOVABLE	SENATE RPT 95-1263	SEN RPT 95-1263, 1978-3 CB 315	N/A	ALL BUSINESS	TANGIBLE PERSONAL PROPERTY	
FLOOR REINFORCED-EXCESS COSTS	REVENUE RULING 69-558	REV RUL 69-558, 1969-2 CB 4	N/A	YARNS-FABRICS/YARNS	STRUCTURAL COMPONENT	
FLOORING-SANITARY TILING	MORRISON	90-1 USTC 50,034	11 TH	RESTAURANTS	STRUCTURAL COMPONENT	
FLOOR-INSULATED COOLER	MORRISON	90-1 USTC 50,034	11 TH	RESTAURANTS	STRUCTURAL COMPONENT	
FLOOR-INSULATED FREEZER	MORRISON	90-1 USTC 50,034	11 TH	RESTAURANTS	STRUCTURAL COMPONENT	
FLOOR-RAISED	FSA 200110001				STRUCTURAL COMPONENT	ENCOMPASSED 90% OF THE FLOOR SPACE
FLOOR-RAISED	REVENUE RULING 74-391	REV RUL 74-391, 1974-2 CB 9	N/A	MANUFACTURING	TANGIBLE PERSONAL PROPERTY	SPECIFIC USE FOR COMPUTERS
FLOORS	TREAS. REG. 1.48-1	TREAS. REG. 1.48-1	N/A	ALL BUSINESS	STRUCTURAL COMPONENT	
FLOOR-WOOD BLOCK	REVENUE RULING 74-391	REV RUL 74-391, 1974-2 CB 9	N/A	MANUFACTURING	STRUCTURAL COMPONENT	INSTALLED WITH MASTIC TO CONCRETE SUBFLOOR
FLUORESCENT FIXTURES	CENTRAL CITRUS	58 TC 37	9 TH	FOOD MANUFACTURER	TANGIBLE PERSONAL PROPERTY	RELATES TO EQUIPMENT USED IN MANUFACTURING
FOLDING WALL PARTITIONS	HOSPITAL CORP OF AMERICA	109 TC 21	6TH	HOSPITAL	TANGIBLE PERSONAL PROPERTY	
FOOTINGS-GASOLINE STATION PUMP CANOPIES	REVENUE RULING 2003-54	REV RUL 2003-54 IRB 2003-23	N/A	FUEL OUTLET	INHERENTLY PERMANENT STRUCTURE	
FOUNDATION STABILIZATION	AMERICAN BEMBERG CORP	49-2 USTC ¶9460	6TH	MANUFACTURER	EXPENSED	
GAS DISTRIBUTION-UNDERGROUND	REVENUE RULING 80-93	REV RUL 80-93, 1980-1 CB 50	N/A	LESSORS-MOBILE HOME PARK	LAND	NOT OWNED BY THE TAX PAYER
GAS STATIONS	BROOKSHIRE BROTHERS	TC MEMO 2001-150, 81 TCM 1799	5TH	RETAILER	INHERENTLY PERMANENT STRUCTURE	
GASOLINE PUMP CANOPIES	REVENUE RULING 2003-54	REV RUL 2003-54 IRB 2003-23	N/A	FUEL OUTLET	TANGIBLE PERSONAL PROPERTY	
GOLF GREEN	REVENUE RULING 2001-60				INHERENTLY PERMANENT STRUCTURE	RELATES TO CERTAIN COMPONENTS OF THE GREENS
GRADING-SITE WORK	REVENUE RULING 80-93	REV RUL 80-93, 1980-1 CB 50	N/A	LESSORS-MOBILE HOME PARK	LAND	PREPARES LAND FOR ITS INTENDED USE
GRADING-SITE WORK	EASTWOOD MALL	95-1 USTC 50,236	6TH	MALL OPERATOR	LAND	CREATION OF USABLE LAND
GREASE TRAP	LA PETITE ACADEMY	95-1 USTC 50,193	8 TH	CHILD CARE CENTERS	STRUCTURAL COMPONENT	

Figure 5.3 continued

ASSET CLASSIFICATION TABLE

ASSET	CASE/RULING	CITE	CIRCUIT	BUSINESS CLASSIFICATION FOR DEPRECIATION PURPOSES	CLASSIFICATION	NOTES
GYPSUM WALL	METRO NATIONAL CORPORATION	TC MEMO 1987-38	5TH	LESSORS OF REAL PROPERTY	BOTH	CASE DEALT WITH TWO DIFFERENT BUILDING TYPES-TAXPAYERS PRACTICES WERE CONSIDERED
HAND SINK-KITCHEN	MORRISON	90-1 USTC 50,034	11 TH	RESTAURANTS	STRUCTURAL COMPONENT	
HANGING CHANDELIER	LETTER RULING 8002015	LETTER RULING 8002015, OCTOBER 1, 1979	N/A	RESTAURANT	STRUCTURAL COMPONENT	PROVIDED GENERAL BUILDING LIGHTING
HANGING LANTERN	SHONEYS SOUTH	TC MEMO 1984-413	6TH	RESTAURANT	TANGIBLE PERSONAL PROPERTY	
HEAT DETECTORS	LA PETITE ACADEMY	95-1 USTC 50,193	8 TH	CHILD CARE CENTERS	STRUCTURAL COMPONENT	
HEATERS-UNIT HEATERS	LETTER RULING 7102296400A	LET RUL 7102296400A, FEBRUARY 26, 1971	N/A	MANUFACTURER/WAREHOUSING	OTHER PROPERTY	IF SOLE JUSTIFICATION TEST MET
HEATING/AIR CONDITIONING	LETTER RULING 8848039	LET RUL 8848039 SEPTEMBER 12, 1988	N/A	LESSORS OF REAL PROPERTY	STRUCTURAL COMPONENT	
HIGH BAY ROOF	AC MONK & COMPANY	82-2 USTC ¶9551	4TH	TOBACCO PROCESSING	STRUCTURAL COMPONENT	COULD BE REASONABLY ADAPTED TO OTHER USES
HUMIDITY CONTROL EQUIPMENT	REVENUE RULING 69-558	REV RUL 69-558, 1969-2 CB 4	N/A	YARNS-FABRICS/YARNS	OTHER PROPERTY	SOLE JUSTIFICATION TEST MET
HVAC	TREAS. REG. 1.48-1	TREAS. REG. 1.48-1	N/A	ALL BUSINESS	STRUCTURAL COMPONENT	
HVAC DUCTS	TREAS. REG. 1.48-1	TREAS. REG. 1.48-1	N/A	ALL BUSINESS	STRUCTURAL COMPONENT	
HVAC PIPING	TREAS. REG. 1.48-1	TREAS. REG. 1.48-1	N/A	ALL BUSINESS	STRUCTURAL COMPONENT	
HVAC-COMPRESSOR	TREAS. REG. 1.48-1	TREAS. REG. 1.48-1	N/A	ALL BUSINESS	STRUCTURAL COMPONENT	
HVAC-MOTOR	TREAS. REG. 1.48-1	TREAS. REG. 1.48-1	N/A	ALL BUSINESS	STRUCTURAL COMPONENT	
IDENTITY SYMBOLS	SENATE RPT 95-1263	SEN RPT 95-1263, 1978-3 CB 315	N/A	ALL BUSINESS	TANGIBLE PERSONAL PROPERTY	
INDOOR STORAGE FACILITIES	LETTER RULING 7102296400A	LET RUL 7102296400A, FEBRUARY 26, 1971	N/A	MANUFACTURER/WAREHOUSING	STRUCTURAL COMPONENT	
INSULATED FLOORING	MORRISON	90-1 USTC 50,034	11 TH	RESTAURANTS	STRUCTURAL COMPONENT	
KITCHEN STEAM LINES	HOSPITAL CORP OF AMERICA	109 TC 21	6TH	HOSPITAL	TANGIBLE PERSONAL PROPERTY	
KITCHEN DOORS	LA PETITE ACADEMY	95-1 USTC 50,193	8 TH	CHILD CARE CENTERS	STRUCTURAL COMPONENT	
KITCHEN EXHAUST	HOSPITAL CORP OF AMERICA	109 TC 21	6TH	HOSPITAL	TANGIBLE PERSONAL PROPERTY	
KITCHEN HAND SINK	MORRISON	90-1 USTC 50,034	11 TH	RESTAURANTS	STRUCTURAL COMPONENT	
KITCHEN HOOD	HOSPITAL CORP OF AMERICA	109 TC 21	6TH	HOSPITAL	TANGIBLE PERSONAL PROPERTY	
KITCHEN HOOK-UPS	FSA 200203009	FSA 200203009	N/A	CASINO	TANGIBLE PERSONAL PROPERTY	
LAMINATED PLASTICS-CABINETS	METRO NATIONAL CORPORATION	TC MEMO 1987-38	5TH	LESSORS OF REAL PROPERTY	TANGIBLE PERSONAL PROPERTY	

Figure 5.3 continued

ASSET CLASSIFICATION TABLE

ASSET	CASE/RULING	CITE	CIRCUIT	BUSINESS CLASSIFICATION FOR DEPRECIATION PURPOSES	CLASSIFICATION	NOTES
LAND PREPARATION	REVENUE RULING 65-265	REV RUL 65-265, 1965-2 CB 52	N/A	PROCESSING/STORAGE	LAND	UNTIL READY FOR INTENDED USE
LAND PREPARATION COSTS	REVENUE RULING 80-93	REV RUL 80-93, 1980-1 CB 50	N/A	LESSORS-MOBILE HOME PARK	INHERENTLY PERMANENT STRUCTURE	ASSOCIATED WITH OTHER ASSETS WITH A DETERMINABLE USEFUL LIFE
LAND PREPARATION-SETTING OF BUILDINGS	REVENUE RULING 65-265	REV RUL 65-265, 1965-2 CB 52	N/A	PROCESSING/STORAGE	STRUCTURAL COMPONENT	SINCE IT RELATES TO THE BUILDINGS
LAND PREPARATION-SIDEWALKS/ROAD WAYS	REVENUE RULING 65-265	REV RUL 65-265, 1965-2 CB 52	N/A	PROCESSING/STORAGE	INHERENTLY PERMANENT STRUCTURE	RELATES TO THE ASSETS BEING CONSTRUCTED
LANDSCAPING	LETTER RULING 8848039	LET RUL 8848039 SEPTEMBER 12, 1988	N/A	LESSORS OF REAL PROPERTY	INHERENTLY PERMANENT STRUCTURE	
LANDSCAPING SHRUBBERY	REVENUE PROCEDURE 87-56	REV PROC 87-56, 1987-2 CB 674	N/A	ALL BUSINESS	INHERENTLY PERMANENT STRUCTURE	
LANDSCAPING	REVENUE PROCEDURE 87-56	REV PROC 87-56, 1987-2 CB 674	N/A	ALL BUSINESS	INHERENTLY PERMANENT STRUCTURE	
LAY-IN LIGHTING	METRO NATIONAL CORPORATION	TC MEMO 1987-38	5TH	LESSORS OF REAL PROPERTY	STRUCTURAL COMPONENT	
LIGHTING FIXTURES	TREAS. REG. 1.48-1	TREAS. REG. 1.48-1	N/A	ALL BUSINESS	STRUCTURAL COMPONENT	
LIGHTING-ACCENT DECORATIVE LIGHTING	METRO NATIONAL CORPORATION	TC MEMO 1987-38	5TH	LESSORS OF REAL PROPERTY	TANGIBLE PERSONAL PROPERTY	
LIGHTING-EXTERIOR DECORATIVE	METRO NATIONAL CORPORATION	TC MEMO 1987-38	5TH	LESSORS OF REAL PROPERTY	TANGIBLE PERSONAL PROPERTY	SPECIAL LIGHTING
LIGHTING-EXTERIOR SECURITY	METRO NATIONAL CORPORATION	TC MEMO 1987-38	5TH	LESSORS OF REAL PROPERTY	TANGIBLE PERSONAL PROPERTY	
LIGHTING-GROW LIGHTS	METRO NATIONAL CORPORATION	TC MEMO 1987-38	5TH	LESSORS OF REAL PROPERTY	TANGIBLE PERSONAL PROPERTY	
LIGHTING-LAY IN PANELS	METRO NATIONAL CORPORATION	TC MEMO 1987-38	5TH	LESSORS OF REAL PROPERTY	STRUCTURAL COMPONENT	
LIGHTING-LOADING DOCKS	CONSOLIDATED FREIGHTWAYS	83-1 USTC 9420		TRANSPORTATION/WAREHOUSE	STRUCTURAL COMPONENT	
LIGHTS-OVER BED	HOSPITAL CORP OF AMERICA	109 TC 21	6TH	HOSPITAL	STRUCTURAL COMPONENT	
LOADING DOCKS	CONSOLIDATED FREIGHTWAYS	83-1 USTC 9420		TRANSPORTATION/WAREHOUSE	STRUCTURAL COMPONENT	PART OF THE BUILDING
MANSARD ROOF	LA PETITE ACADEMY	95-1 USTC 50,193	8 TH	CHILD CARE CENTERS	STRUCTURAL COMPONENT	
MILLWORK-CROWN MOLDINGS	FSA 200203009	FSA 200203009	N/A	CASINO	TANGIBLE PERSONAL PROPERTY	
MILLWORK-LATTICE CEILINGS	FSA 200203009	FSA 200203009	N/A	CASINO	TANGIBLE PERSONAL PROPERTY	
MILLWORK-LATTICE WALLS	FSA 200203009	FSA 200203009	N/A	CASINO	TANGIBLE PERSONAL PROPERTY	
MILLWORK-ORNATE WALL PANELING	FSA 200203009	FSA 200203009	N/A	CASINO	TANGIBLE PERSONAL PROPERTY	

Figure 5.3 continued

ASSET CLASSIFICATION TABLE

ASSET	CASE/RULING	CITE	CIRCUIT	BUSINESS CLASSIFICATION FOR DEPRECIATION PURPOSES	CLASSIFICATION	NOTES
MOTOR CONTROL CENTERS	AC MONK & COMPANY	82-2 USTC ¶9551	4TH	TOBACCO PROCESSING	BOTH	COULD BE REASONABLY ADAPTED TO OTHER USES- REMANDED TO LOWER COURT FOR ITEM BY ITEM DETERMINATION
MOTOR CONTROL CENTERS	ILLINOIS CEREAL MILLS	86-1 USTC ¶9371	7TH	CEREAL MANUFACTURER	OTHER PROPERTY	RELATES TO MANUFACTURING EQUIPMENT
MOTORS-AIR HANDLING	REVENUE RULING 69-558	REV RUL 69-558, 1969-2 CB 4	N/A	YARNS-FABRICS/YARNS	OTHER PROPERTY	SOLE JUSTIFICATION TEST MET
MOVABLE PARTITION	SENATE RPT 95-1263	SEN RPT 95-1263, 1978-3 CB 315	N/A	ALL BUSINESS	TANGIBLE PERSONAL PROPERTY	
NIGHT DEPOSIT FACILITY	REVENUE RULING 65-79			BANK	TANGIBLE PERSONAL PROPERTY	
ORNAMENTAL FIXTURES	SENATE RPT 95-1263	SEN RPT 95-1263, 1978-3 CB 315	N/A	ALL BUSINESS	TANGIBLE PERSONAL PROPERTY	
OUTDOOR STORAGE PADS	LETTER RULING 7102296400A	LET RUL 7102296400A, FEBRUARY 26, 1971	N/A	MANUFACTURER/WAREHOUSING	OTHER PROPERTY	USED PREDOMINANTLY IN TRANSPORT OF RAW MATERIALS INCLUDED IN MANUFACTURING CLASSIFICATION
OVERHEAD DOORS	CONSOLIDATED FREIGHTWAYS	83-1 USTC 9420		TRANSPORTATION/WAREHOUSE	STRUCTURAL COMPONENT	
PADS-OUTDOOR STORAGE	LETTER RULING 7102296400A	LET RUL 7102296400A, FEBRUARY 26, 1971	N/A	MANUFACTURER/WAREHOUSING	OTHER PROPERTY	USED PREDOMINANTLY IN TRANSPORT OF RAW MATERIALS INCLUDED IN MANUFACTURING CLASSIFICATION
PAPER TOWEL DISPENSER	LA PETITE ACADEMY	95-1 USTC 50,193	8 TH	CHILD CARE CENTERS	STRUCTURAL COMPONENT	
PARTITION MOVABLE	SENATE RPT 95-1263	SEN RPT 95-1263, 1978-3 CB 315	N/A	ALL BUSINESS	TANGIBLE PERSONAL PROPERTY	
PARTITION REMOVABLE	SENATE RPT 95-1263	SEN RPT 95-1263, 1978-3 CB 315	N/A	ALL BUSINESS	TANGIBLE PERSONAL PROPERTY	
PARTITIONS	METRO NATIONAL CORPORATION	TC MEMO 1987-38	5TH	LESSORS OF REAL PROPERTY	BOTH	CASE DEALT WITH TWO DIFFERENT BUILDING TYPES-TAXPAYERS PRACTICES WERE CONSIDERED
PARTITIONS	TREAS. REG. 1.48-1	TREAS. REG. 1.48-1	N/A	ALL BUSINESS	STRUCTURAL COMPONENT	
PARTITIONS-CORRIDORS	METRO NATIONAL CORPORATION	TC MEMO 1987-38	5TH	LESSORS OF REAL PROPERTY	TANGIBLE PERSONAL PROPERTY	MAY BE MOVED TO SERVE THE NEEDS OF TENANTS.
PARTITIONS-STORE FRONT	METRO NATIONAL CORPORATION	TC MEMO 1987-38	5TH	LESSORS OF REAL PROPERTY	STRUCTURAL COMPONENT	REQUIRED TO SEPARATE STORE FROM MALL AREA
PARTITIONS-TOILET	METRO NATIONAL CORPORATION	TC MEMO 1987-38	5TH	LESSORS OF REAL PROPERTY	STRUCTURAL COMPONENT	NO OTHER USE FOR THIS ITEM "AS LONG AS THERE ARE TOILETS THERE WILL BE PARTITIONS"
PATIENT HANDRAILS	HOSPITAL CORP OF AMERICA	109 TC 21	6TH	HOSPITAL	TANGIBLE PERSONAL PROPERTY	
PERIPHERAL EQUIPMENT-COMPUTERS-GEN BUSINESS	REVENUE PROCEDURE 87-56	REV PROC 87-56, 1987-2 CB 674	N/A	ALL BUSINESS	TANGIBLE PERSONAL PROPERTY	
PERMANENT FLOOR COVERING	TREAS. REG. 1.48-1	TREAS. REG. 1.48-1	N/A	ALL BUSINESS	STRUCTURAL COMPONENT	

Figure 5.3 continued

ASSET CLASSIFICATION TABLE

ASSET	CASE/RULING	CITE	CIRCUIT	BUSINESS CLASSIFICATION FOR DEPRECIATION PURPOSES	CLASSIFICATION	NOTES
PERMANENT WALL COVERING	TREAS. REG. 1.48-1	TREAS. REG. 1.48-1	N/A	ALL BUSINESS	STRUCTURAL COMPONENT	
PLAYGROUND EQUIPMENT	LETTER RULING 8848039	LET RUL 8848039 SEPTEMBER 12, 1988	N/A	LESSORS OF REAL PROPERTY	INHERENTLY PERMANENT STRUCTURE	
PLAYGROUND FENCING	LA PETITE ACADEMY	95-1 USTC 50,193	8 TH	CHILD CARE CENTERS	INHERENTLY PERMANENT STRUCTURE	
PLUMBING	TREAS. REG. 1.48-1	TREAS. REG. 1.48-1	N/A	ALL BUSINESS	STRUCTURAL COMPONENT	
PLUMBING FIXTURES	TREAS. REG. 1.48-1	TREAS. REG. 1.48-1	N/A	ALL BUSINESS	STRUCTURAL COMPONENT	
PLUMBING SPECIAL CONNECTION	REVENUE RULING 66-299	REV RUL 66-299, 1966-2 CB 14	N/A	MANUFACTURER		RELATES TO MANUFACTURING EQUIPMENT
PLUMBING-EQUIPMENT	HOSPITAL CORP OF AMERICA	109 TC 21	6TH	HOSPITAL	TANGIBLE PERSONAL PROPERTY	
PLUMBING-SPECIALTY X-RAY	HOSPITAL CORP OF AMERICA	109 TC 21	6TH	HOSPITAL	TANGIBLE PERSONAL PROPERTY	
POWER PANELS	ILLINOIS CEREAL MILLS	86-1 USTC ¶9371	7TH	CEREAL MANUFACTURER	OTHER PROPERTY	RELATES TO MANUFACTURING EQUIPMENT
POWER SUPPLY DISTRIBUTION	LETTER RULING 7102296400A	LET RUL 7102296400A, FEBRUARY 26, 1971	N/A	MANUFACTURER/WAREHOUSING	OTHER PROPERTY	RELATES TO INDIVIDUAL MANUFACTURING ASSETS
PRESS PIT	AC MONK & COMPANY	82-2 USTC ¶9551	4TH	TOBACCO PROCESSING	OTHER PROPERTY	CASE NOTES THE IRS ALLOWED CREDIT
PRIMARY ELECTRICAL SYSTEM	SCOTT PAPER	74 TC 137		MANUFACTURER-PAPER PRODUCTS	STRUCTURAL COMPONENT	FUNCTIONAL ALLOCATION USED
PRIMARY ELECTRICAL SYSTEM	SCOTT PAPER	74 TC 137		MANUFACTURER-PAPER PRODUCTS	TANGIBLE PERSONAL PROPERTY	FUNCTIONAL ALLOCATION USED
PUMPS-AIR HANDLING	REVENUE RULING 69-558	REV RUL 69-558, 1969-2 CB 4	N/A	YARNS-FABRICS/YARNS	OTHER PROPERTY	SOLE JUSTIFICATION TEST MET
PYLON SIGN	FSA 200203009	FSA 200203009	N/A	CASINO	INHERENTLY PERMANENT STRUCTURE	
PYLON SIGN-MESSAGE BOARD	FSA 200203009	FSA 200203009	N/A	CASINO	TANGIBLE PERSONAL PROPERTY	
RADIO TOWERS	REVENUE PROCEDURE 87-56	REV PROC 87-56, 1987-2 CB 674	N/A	ALL BUSINESS	INHERENTLY PERMANENT STRUCTURE	
RAILROAD DOCK-RAISED CONCRETE	AC MONK & COMPANY	82-2 USTC ¶9551	4TH	TOBACCO PROCESSING	STRUCTURAL COMPONENT	COULD BE REASONABLY ADAPTED TO OTHER USES
RAISED FLOORING-COMPUTER INSTALLATION	LETTER RULING 7102296400A	LET RUL 7102296400A, FEBRUARY 26, 1971	N/A	MANUFACTURER/WAREHOUSING	TANGIBLE PERSONAL PROPERTY	INSTALLED IN ADDITION TO THE GENERAL FLOORING
RAISED ROOF	AC MONK & COMPANY	82-2 USTC ¶9551	4TH	TOBACCO PROCESSING	STRUCTURAL COMPONENT	COULD BE REASONABLY ADAPTED TO OTHER USES
RECESSED LIGHTING	LETTER RULING 8002015	LETTER RULING 8002015, OCTOBER 1, 1979	N/A	RESTAURANT	STRUCTURAL COMPONENT	PROVIDED GENERAL BUILDING LIGHTING
RECORD VAULT	REVENUE RULING 65-79			BANK	TANGIBLE PERSONAL PROPERTY	
REINFORCED FLOORING	REVENUE RULING 69-558	REV RUL 69-558, 1969-2 CB 4	N/A	YARNS-FABRICS/YARNS	STRUCTURAL COMPONENT	

Figure 5.3 continued

ASSET CLASSIFICATION TABLE

ASSET	CASE/RULING	CITE	CIRCUIT	BUSINESS CLASSIFICATION FOR DEPRECIATION PURPOSES	CLASSIFICATION	NOTES
REMOVABLE PARTITION	SENATE RPT 95-1263	SEN RPT 95-1263, 1978-3 CB 315	N/A	ALL BUSINESS	TANGIBLE PERSONAL PROPERTY	
RESEARCH ROADWAYS	LETTER RULING 7102296400A	LET RUL 7102296400A, FEBRUARY 26, 1971	N/A	MANUFACTURER/WAREHOUSING	OTHER PROPERTY	USED IN TESTING WHICH IS PART OF THE MANUFACTURING PROCESS
RETAINING WALLS-RAW MATERIAL STORAGE	LETTER RULING 7102296400A	LET RUL 7102296400A, FEBRUARY 26, 1971	N/A	MANUFACTURER/WAREHOUSING	OTHER PROPERTY	RELATES TO PRODUCTION ACTIVITY
REVETMENTS	REVENUE RULING 77-270	REV RUL 77-270, 1977-2 CB 79	N/A	STEEL PRODUCER	LAND	ASSOCIATED WITH THE CREATION OF LAND
ROADS	REVENUE PROCEDURE 87-56	REV PROC 87-56, 1987-2 CB 674	N/A	ALL BUSINESS	INHERENTLY PERMANENT STRUCTURE	
ROADWAYS-RESEARCH	LETTER RULING 7102296400A	LET RUL 7102296400A, FEBRUARY 26, 1971	N/A	MANUFACTURER/WAREHOUSING	OTHER PROPERTY	USED IN TESTING WHICH IS PART OF THE MANUFACTURING PROCESS
ROADWAYS-TRUCKS	LETTER RULING 7102296400A	LET RUL 7102296400A, FEBRUARY 26, 1971	N/A	MANUFACTURER/WAREHOUSING	OTHER PROPERTY	USED PREDOMINANTLY IN TRANSPORT OF RAW MATERIALS INCLUDED IN MANUFACTURING CLASSIFICATION
ROLLING DOORS-TRAIN ENTRANCE	LETTER RULING 7102296400A	LET RUL 7102296400A, FEBRUARY 26, 1971	N/A	MANUFACTURER/WAREHOUSING	STRUCTURAL COMPONENT	
ROLLING DOORS-TRUCK ENTRANCE	LETTER RULING 7102296400A	LET RUL 7102296400A, FEBRUARY 26, 1971	N/A	MANUFACTURER/WAREHOUSING	STRUCTURAL COMPONENT	
ROOF REPAIR	OBERMAN MANUFACTURING	47 TC 471		MANUFACTURING	EXPENSED	
ROOF-EXCESS COSTS	REVENUE RULING 69-558	REV RUL 69-558, 1969-2 CB 4	N/A	YARNS-FABRICS/YARNS	STRUCTURAL COMPONENT	
ROUGH GRADING-SITE WORK	REVENUE RULING 80-93	REV RUL 80-93, 1980-1 CB 50	N/A	LESSORS-MOBILE HOME PARK	LAND	PREPARES LAND FOR ITS INTENDED USE
SAFES	REVENUE PROCEDURE 87-56	REV PROC 87-56, 1987-2 CB 674	N/A	ALL BUSINESS	TANGIBLE PERSONAL PROPERTY	
SAFETY EQUIPMENT	LETTER RULING 7102296400A	LET RUL 7102296400A, FEBRUARY 26, 1971	N/A	MANUFACTURER/WAREHOUSING	OTHER PROPERTY	LOCALIZED IN A FACTORY OR RELATED TO SPECIFIC MACHINERY OR PROCESSES
SANITARY TILING-FLOORS	MORRISON	90-1 USTC 50,034	11 TH	RESTAURANTS	STRUCTURAL COMPONENT	
SANITARY TILING-WALLS	MORRISON	90-1 USTC 50,034	11 TH	RESTAURANTS	STRUCTURAL COMPONENT	
SCALE PITS	LETTER RULING 7102296400A	LET RUL 7102296400A, FEBRUARY 26, 1971	N/A	MANUFACTURER/WAREHOUSING	OTHER PROPERTY	
SECURITY LIGHTING	METRO NATIONAL CORPORATION	TC MEMO 1987-38	5TH	LESSORS OF REAL PROPERTY	TANGIBLE PERSONAL PROPERTY	
SEWERS	REVENUE PROCEDURE 87-56	REV PROC 87-56, 1987-2 CB 674	N/A	ALL BUSINESS	INHERENTLY PERMANENT STRUCTURE	
SHEDS	LETTER RULING 7102296400A	LET RUL 7102296400A, FEBRUARY 26, 1971	N/A	MANUFACTURER/WAREHOUSING	INHERENTLY PERMANENT STRUCTURE	
SIDEWALKS	REVENUE PROCEDURE 87-56	REV PROC 87-56, 1987-2 CB 674	N/A	ALL BUSINESS	INHERENTLY PERMANENT STRUCTURE	
SIDEWALKS	REVENUE RULING 80-93	REV RUL 80-93, 1980-1 CB 50	N/A	LESSORS-MOBILE HOME PARK	INHERENTLY PERMANENT STRUCTURE	
SINKS	TREAS. REG. 1.48-1	TREAS. REG. 1.48-1	N/A	ALL BUSINESS	STRUCTURAL COMPONENT	

Figure 5.3 continued

ASSET CLASSIFICATION TABLE

ASSET	CASE/RULING	CITE	CIRCUIT	BUSINESS CLASSIFICATION FOR DEPRECIATION PURPOSES	CLASSIFICATION	NOTES
SITE UTILITIES	FSA 200203009	FSA 200203009	N/A	CASINO	INHERENTLY PERMANENT STRUCTURE	
SITE WORK-CONCRETE DRIVEWAYS	REVENUE RULING 80-93	REV RUL 80-93, 1980-1 CB 50	N/A	LESSORS-MOBILE HOME PARK	INHERENTLY PERMANENT STRUCTURE	
SITE WORK-LAUNDRY FACILITIES	REVENUE RULING 80-93	REV RUL 80-93, 1980-1 CB 50	N/A	LESSORS-MOBILE HOME PARK	INHERENTLY PERMANENT STRUCTURE	
SITE WORK-MOBILE HOME SUPPORTS	REVENUE RULING 80-93	REV RUL 80-93, 1980-1 CB 50	N/A	LESSORS-MOBILE HOME PARK	INHERENTLY PERMANENT STRUCTURE	
SITE WORK-PATIO	REVENUE RULING 80-93	REV RUL 80-93, 1980-1 CB 50	N/A	LESSORS-MOBILE HOME PARK	INHERENTLY PERMANENT STRUCTURE	
SITE WORK-STORM SEWERS	REVENUE RULING 80-93	REV RUL 80-93, 1980-1 CB 50	N/A	LESSORS-MOBILE HOME PARK	INHERENTLY PERMANENT STRUCTURE	
SMOKE DETECTORS	LA PETITE ACADEMY	95-1 USTC 50,193	8 TH	CHILD CARE CENTERS	STRUCTURAL COMPONENT	
SMOKE DETECTORS	LETTER RULING 8848039	LET RUL 8848039 SEPTEMBER 12, 1988	N/A	LESSORS OF REAL PROPERTY	STRUCTURAL COMPONENT	
SNOWMAKING SYSTEM DISTRIBUTION PIPES AND WIRING	REVENUE RULING 85-93	REV RUL 85-93, 1985-2 CB 9	N/A	SKI RESORT OPERATOR	INHERENTLY PERMANENT STRUCTURE	
SNOWMAKING SYSTEM-GUNS AND OTHER EQUIPMENT	REVENUE RULING 85-93	REV RUL 85-93, 1985-2 CB 9	N/A	SKI RESORT OPERATOR	TANGIBLE PERSONAL PROPERTY	
SNOWMAKING SYSTEM-PUMP HOUSE	REVENUE RULING 85-93	REV RUL 85-93, 1985-2 CB 9	N/A	SKI RESORT OPERATOR	STRUCTURAL COMPONENT	
SPECIAL LIGHTING	SENATE RPT 95-1263	SEN RPT 95-1263, 1978-3 CB 315	N/A	ALL BUSINESS	TANGIBLE PERSONAL PROPERTY	
SPOTLIGHTS	CENTRAL CITRUS	58 TC 37	9 TH	FOOD MANUFACTURER	TANGIBLE PERSONAL PROPERTY	RELATES TO EQUIPMENT USED IN MANUFACTURING
SPRINKLER HEADS-OUTDOOR	METRO NATIONAL CORPORATION	TC MEMO 1987-38	5TH	LESSORS OF REAL PROPERTY	INHERENTLY PERMANENT STRUCTURE	PART OF AND GIVES FUNCTION TO THE OUTDOOR SPRINKLER SYSTEM
SPRINKLER SYSTEMS	TREAS. REG. 1.48-1	TREAS. REG. 1.48-1	N/A	ALL BUSINESS	STRUCTURAL COMPONENT	
STAIRS	TREAS. REG. 1.48-1	TREAS. REG. 1.48-1	N/A	ALL BUSINESS	STRUCTURAL COMPONENT	
STORAGE ROOM	AC MONK & COMPANY	82-2 USTC ¶9551	4TH	TOBACCO PROCESSING	STRUCTURAL COMPONENT	COULD BE REASONABLY ADAPTED TO OTHER USES
STREETS	LETTER RULING 8848039	LET RUL 8848039 SEPTEMBER 12, 1988	N/A	LESSORS OF REAL PROPERTY	INHERENTLY PERMANENT STRUCTURE	
SWITCHBOARDS	ILLINOIS CEREAL MILLS	86-1 USTC ¶9371	7TH	CEREAL MANUFACTURER	OTHER PROPERTY	RELATES TO MANUFACTURING EQUIPMENT
SWITCHES 10 AMP OR 30 AMP	CENTRAL CITRUS	58 TC 37	9 TH	FOOD MANUFACTURER	TANGIBLE PERSONAL PROPERTY	RELATES TO EQUIPMENT USED IN MANUFACTURING

Figure 5.3 continued

ASSET CLASSIFICATION TABLE

ASSET	CASE/RULING	CITE	CIRCUIT	BUSINESS CLASSIFICATION FOR DEPRECIATION PURPOSES	CLASSIFICATION	NOTES
SWITCHGEAR PANEL BOARDS	AC MONK & COMPANY	82-2 USTC ¶9551	4TH	TOBACCO PROCESSING	BOTH	COULD BE REASONABLY ADAPTED TO OTHER USES- REMANDED TO LOWER COURT FOR ITEM BY ITEM DETERMINATION
TELEPHONE POLE REMOVAL	REVENUE RULING 2000-7	REV RUL 2000-7, 2000-1 CB 712	N/A	TELEPHONE COMPANY	EXPENSED	COST OF RETIRING OLD ASSET
TELEVISION TOWERS	REVENUE PROCEDURE 87-56	REV PROC 87-56, 1987-2 CB 674	N/A	ALL BUSINESS	INHERENTLY PERMANENT STRUCTURE	
TOILET CABINET	LA PETITE ACADEMY	95-1 USTC 50,193	8 TH	CHILD CARE CENTERS	STRUCTURAL COMPONENT	
TOILET PAPER DISPENSER	LA PETITE ACADEMY	95-1 USTC 50,193	8 TH	CHILD CARE CENTERS	STRUCTURAL COMPONENT	
TOILET PARTITIONS	METRO NATIONAL CORPORATION	TC MEMO 1987-38	5TH	LESSORS OF REAL PROPERTY	STRUCTURAL COMPONENT	NO OTHER USE FOR THIS ITEM "AS LONG AS THERE ARE TOILETS THERE WILL BE PARTITIONS"
TRANSFORMERS	AC MONK & COMPANY	82-2 USTC ¶9551	4TH	TOBACCO PROCESSING	BOTH	COULD BE REASONABLY ADAPTED TO OTHER USES- REMANDED TO LOWER COURT FOR ITEM BY ITEM DETERMINATION
TRANSFORMERS	ILLINOIS CEREAL MILLS	86-1 USTC ¶9371	7TH	CEREAL MANUFACTURER	OTHER PROPERTY	RELATES TO MANUFACTURING EQUIPMENT
TRENCHING-BUILDING FOUNDATION	EASTWOOD MALL	95-1 USTC 50,236	6TH	MALL OPERATOR	STRUCTURAL COMPONENT	
TRENCHING-UTILITIES	EASTWOOD MALL	95-1 USTC 50,236	6TH	MALL OPERATOR	STRUCTURAL COMPONENT	
TRUCK ROADWAYS	LETTER RULING 7102296400A	LET RUL 7102296400A, FEBRUARY 26, 1971	N/A	MANUFACTURER/WAREHOUSING	OTHER PROPERTY	USED PREDOMINANTLY IN TRANSPORT OF RAW MATERIALS INCLUDED IN MANUFACTURING CLASSIFICATION
VINYL FLOOR COVERINGS	HOSPITAL CORP OF AMERICA	109 TC 21	6TH	HOSPITAL	TANGIBLE PERSONAL PROPERTY	
VINYL WALL COVERINGS	HOSPITAL CORP OF AMERICA	109 TC 21	6TH	HOSPITAL	TANGIBLE PERSONAL PROPERTY	
WALK UP WINDOW	REVENUE RULING 65-79			BANK	TANGIBLE PERSONAL PROPERTY	
WALKS	LETTER RULING 8848039	LET RUL 8848039 SEPTEMBER 12, 1988	N/A	LESSORS OF REAL PROPERTY	INHERENTLY PERMANENT STRUCTURE	
WALL COVERINGS	FSA 200203009	FSA 200203009	N/A	CASINO	TANGIBLE PERSONAL PROPERTY	
WALL PANEL INSERTS-CONDIMENT HOLDERS	SENATE RPT 95-1263	SEN RPT 95-1263, 1978-3 CB 315	N/A	ALL BUSINESS	TANGIBLE PERSONAL PROPERTY	
WALL PANEL INSERTS-PICTURE FRAMES	SENATE RPT 95-1263	SEN RPT 95-1263, 1978-3 CB 315	N/A	ALL BUSINESS	TANGIBLE PERSONAL PROPERTY	
WALL TO WALL CARPETING	REVENUE RULING 67-349	REV RUL 67-349 1967-2 CB 48	N/A	MOTEL OPERATOR	TANGIBLE PERSONAL PROPERTY	
WALLS	TREAS. REG. 1.48-1	TREAS. REG. 1.48-1	N/A	ALL BUSINESS	STRUCTURAL COMPONENT	

Figure 5.3 continued

ASSET CLASSIFICATION TABLE

ASSET	CASE/RULING	CITE	CIRCUIT	BUSINESS CLASSIFICATION FOR DEPRECIATION PURPOSES	CLASSIFICATION	NOTES
WALLS-EXCESS COSTS	REVENUE RULING 69-558	REV RUL 69-558, 1969-2 CB 4	N/A	YARNS-FABRICS/YARNS	STRUCTURAL COMPONENT	
WALLS-SANITARY TILING	MORRISON	90-1 USTC 50,034	11 TH	RESTAURANTS	STRUCTURAL COMPONENT	
WASTEWATER TREATMENT FACILITY	FSA 002230	FSA 002230, DECEMBER 2, 1997	N/A	TEXTILE MANUFACTURER	INHERENTLY PERMANENT STRUCTURE	ASSET CLASS 00.3
WATER DISTRIBUTION SYSTEM	LETTER RULING 8848039	LET RUL 8848039 SEPTEMBER 12, 1988	N/A	LESSORS OF REAL PROPERTY	INHERENTLY PERMANENT STRUCTURE	
WATER DISTRIBUTION SYSTEM	LETTER RULING 7102296400A	LET RUL 7102296400A, FEBRUARY 26, 1971	N/A	MANUFACTURER/WAREHOUSING	OTHER PROPERTY	RELATES TO INDIVIDUAL MANUFACTURING ASSETS
WATER HEATERS	LETTER RULING 8848039	LET RUL 8848039 SEPTEMBER 12, 1988	N/A	LESSORS OF REAL PROPERTY	STRUCTURAL COMPONENT	
WATER TANKS-AIR HANDLING	REVENUE RULING 69-558	REV RUL 69-558, 1969-2 CB 4	N/A	YARNS-FABRICS/YARNS	OTHER PROPERTY	SOLE JUSTIFICATION TEST MET
WATERWAYS	REVENUE PROCEDURE 87-56	REV PROC 87-56, 1987-2 CB 674	N/A	ALL BUSINESS	INHERENTLY PERMANENT STRUCTURE	
WHARVES	REVENUE PROCEDURE 87-56	REV PROC 87-56, 1987-2 CB 674	N/A	ALL BUSINESS	INHERENTLY PERMANENT STRUCTURE	
WINDOW TREATMENTS	LETTER RULING 8848039	LET RUL 8848039 SEPTEMBER 12, 1988	N/A	LESSORS OF REAL PROPERTY	TANGIBLE PERSONAL PROPERTY	
WINDOWS	TREAS. REG. 1.48-1	TREAS. REG. 1.48-1	N/A	ALL BUSINESS	STRUCTURAL COMPONENT	
WIRING-TELEPHONE COMMUNICATION EQUIPMENT	HOSPITAL CORP OF AMERICA	109 TC 21	6TH	HOSPITAL	TANGIBLE PERSONAL PROPERTY	
WOOD BLOCK FLOORING	LETTER RULING 7102296400A	LET RUL 7102296400A, FEBRUARY 26, 1971	N/A	MANUFACTURER/WAREHOUSING	TANGIBLE PERSONAL PROPERTY	SHOWN THAT NEEDS OF PRODUCTION REQUIRED INSTALLATION
	REVENUE RULING 70-160				TANGIBLE PERSONAL PROPERTY	
	REVENUE RULING 79-183	1979-1 CB 44				
	REVENUE RULING 79-181	1979-1 CB 41				

Figure 5.3 continued

Chapter 6

Asset Allocation Methodologies

As discussed in Part I, the methodology used to determine the costs of various assets is critical to the accuracy of a cost segregation study. While there is no set procedure for generating a cost segregation study, there are guidelines for each of the different methodologies that can be used to produce more accurate and reliable allocations. At present, there are six industry-accepted approaches for determining the costs of an asset, listed below in order from most to least accurate. All six methods are acknowledged in the 2004 *Cost Segregation Audit Techniques Guide* (published by the IRS for use by its auditors in reviewing cost segregation documentation).

1. Actual cost records
2. Cost estimate replication
3. Survey of project participants
4. Residual estimation
5. Sampling or modeling
6. "Rule of thumb"

Each of the above methodologies has unique attributes that make it suitable to particular applications and circumstances. Some are more accurate than others and therefore establish a more solid basis for asset allocation. The following section will focus on the first three methods, as they have greater accuracy and are best suited to estimating costs. The last three will be discussed in somewhat less detail, because they tend to be less reliable, require more assumptions on the part of the cost segregation practitioner, and can be less favored by the IRS.

Actual Cost Records

This method relies on actual, detailed cost records from the construction of the project. These costs are referred to as *contemporaneous* and are part of the project's accounting record. Sources of information include contracts, subcontracts, change orders, applications for payment, and any separate invoices that were paid by the owner during the construction process. Construction documents, such as the plans and specifications, are also critical in determining the costs of a particular asset, as they constitute the basis for the contract and change orders.

Assuming the cost segregation practitioner has classified the asset correctly (according to Section 1245 or 1250 property life and category), the actual, detailed cost records approach provides the greatest accuracy in establishing asset value. This method is not without its drawbacks, however. Since it is based on actual project cost records and contract documents, it requires that these records be available, intact, and reasonably comprehensive. Unfortunately, only recently completed or ongoing projects tend to meet these criteria. Using the detailed, actual cost records method also requires that the cost segregation practitioner have a certain level of knowledge of the construction process and the ability to read and understand plans and specifications.

Cost Estimate Replication

The second method tends to apply best to recently acquired properties in which the construction cost records are not available or are incomplete. This approach is generally more time-consuming, since it requires that the missing cost information be replicated using published data. This process often requires the skills of a more experienced individual, familiar with construction practices and fluent in both reading and understanding plans. Construction cost estimating skills are another prerequisite.

Cost estimate replications, when detailed and prepared by a qualified individual, should be reasonably accurate. This method is similar to the detailed actual costs approach in that both involve incremental review and costing of the assets. The difference is that the asset's cost is established by estimating its value, rather than using actual cost records. Both methods use construction contract documents, such as the plans and specifications, and (when available) contracts, change orders, invoices, and all other relevant documentation to determine asset allocation.

In most cases, when a property is purchased, plans are not available for use, and construction cost data is not available—all costs need to be recreated or established. Property Appraisal Reports are frequently part of the records in purchase situations.

In the cost estimate replication method, the missing information (actual, detailed costs) is filled in by estimated costs. In some cases, all of the costs will need to be estimated from scratch.

To properly determine an asset's value, it is necessary to rely on an established basis of construction cost data. The most widely recognized and reliable source is published by RSMeans, in both electronic and book form. Means data is organized according to the construction industry's accepted format, which can be referenced for inclusion in the cost segregation study. RSMeans data is also recommended by the IRS in its *Cost Segregation Audit Techniques Guide. (See Chapter 11 for more on cost information sources.)*

Survey of Project Participants

Surveys or letters are sometimes used as a way to acquire supporting information to produce a full set of asset costs. These inquiries are sent to contractors, subcontractors, and vendors to request cost information on specific assets or categories of work performed by those firms. The cost information they provide is then incorporated into the detailed cost or detailed estimate previously outlined.

Getting good information from these surveys requires that the participants have retained project records and that they are willing to respond. Acknowledging that relationships between owners and contractors do not always survive the construction process, some firms may be unwilling to provide any information, accurate or otherwise.

Other Asset Allocation Methods

The remaining three methods are residual estimation, sampling or modeling, and the "rule of thumb" approach. These are acceptable in certain applications, but typically produce somewhat less accurate cost allocations. They are also based on techniques that require cost segregation analysts to make more assumptions about the building or to have a more intimate knowledge of the business or trade housed within. Additionally, these methods do not specifically lend themselves to typical cost engineering practice.

Residual Estimation

This is an abbreviated method for determining the cost of assets that have a depreciation of 5, 7, and 15 years. It is based on deducting the value of the assets from the total project costs, and then assigning all other costs (called the "residual costs") to longer-life assets or to the building structure itself. This method requires that the practitioner, after determining the value of the assets to be recovered, reconcile those costs with the total project cost.

While less time-consuming and detailed, the residual method is also less accurate. Aggressive estimating techniques can produce dramatically different costs for the asset that are sometimes difficult, if not impossible, to reconcile with the total project cost. A word of warning regarding the residual method: auditors are aware that this practice often produces asset costs that are disproportionately in favor of the taxpayer.

Sampling, or Modeling

This method involves creating a model for analyzing identical types of construction and is most popular for redundant-type projects, such as retail stores. The sampling method has gained tremendous popularity among chain stores and restaurants, due to its "modeling" of repeat assets in each location.

The sampling, or modeling, method calls for the practitioner to construct a model of the typical asset based on samples from previous "historical" costs. After the initial outlay of expense for developing the model of the building itself, minor adjustments are made for the time, wage scale, and local cost differences. Separate costs for land development are also adjusted on a location-by-location basis. Done properly, the model can be applied across a wide spectrum as either a percentage or a fixed amount. It avoids having to "re-invent the wheel" each time a study is performed.

Sampling, or modeling, is not without its drawbacks. Incorrect or insufficient "sampling" from either too few or too diverse projects can result in a model that does not accurately reflect typical costs. The practitioner who elects to use this methodology can reasonably be expected to start by explaining in detail how the model was created, should the study be the subject of an audit.

"Rule of Thumb"

This method requires little or no documentation, but practitioners must have sufficient knowledge of the particular industry to approximate the Section 1245 property as a percentage of construction costs. Much of the "data" used in the rule of thumb approach is based on the practitioner's experience with the particular application that is the subject of the study. The rule of thumb method relies on terms like "accepted practice" and "industry standard" to approximate the percentage of personal property that is incorporated in a certain business. The results are often open to different interpretation by different auditors due to the lack of substantiation of project costs. The IRS guidelines caution their auditors as to the lack of accuracy of this particular method.

Each of these last three approaches can lead to a result that relies more on subjective interpretation rather than solid, documented facts and costs. This is because each requires assumptions that are not needed in the detailed cost approaches outlined earlier in this chapter. Cost segregation studies that are developed using one of these three last methodologies do not have the detailed documentation offered by the first three in substantiating a position in the event of an audit.

Site Visit & Photographic Documentation

Independent of the methodology selected, there are several typical requirements of a quality cost segregation study. At the top of that list is the site visit. Whether the property is new or existing construction, the site visit plays an important role in lending credibility to the cost segregation study.

One of the purposes of the site visit is to obtain or verify measurements and quantities needed to estimate assets. When there are no plans or specifications available for a building acquired or constructed in the past, the site visit will be more involved, since it may be one of the few sources of available information. Physical measurements of site improvements, building square footage, structural components, and personal property assets all need to be verified at the site. Asset quality and condition must also be established. Checklists or forms are very helpful in recording this field data.

For new construction, the cost segregation practitioner can gain better insight into the quality and function of the project assets during the site visit. For an existing structure, the site visit allows the analyst to visualize the "wear and tear" or deterioration of the structure as a whole and, in particular, the assets that qualify for accelerated depreciation. Verification that the asset exists and is functioning is also a practical result of the site visit.

A critical part of the site visit is photographic documentation of assets. This helps to substantiate the position taken and is a welcome part of a cost segregation report. For new construction in process, photos of the site development, and even the raw land, are useful in substantiating land development and improvement costs. Photographic documentation can be in the form of still photos or video.

The cost segregation analyst who is performing the study should conduct the site visit, and, as part of his or her narrative report, should include a statement indicating that he or she visited the site and visually witnessed the assets that are the subject of the study. The site visit is most effective if the cost segregation practitioner has had the opportunity to previously review the information provided (any available plans, specifications, cost records, etc.), if the methodology has been selected (at least tentatively), and if the assets have been identified for recovery. This familiarity allows the practitioner to focus on the assets named, and to pursue answers to any questions that may have come up during the review of available information.

Conclusion

Choosing the appropriate methodology to determine and document the costs of assets is crucial to a sound and defensible cost segregation study. Of the six methods outlined in the IRS guidelines, it is important to select the one that best fits the circumstances and demands of the property in question. Some of these techniques, such as detailed cost replication, require construction estimating knowledge and experience, as well as an understanding of construction documentation.

Regardless of the methodology selected, a quality cost segregation study requires familiarity with two distinct professions: tax accounting and construction. Individuals who have expertise and proficiency in all of the skill sets necessary to prepare an entire cost segregation study are rare. For this reason, a quality cost segregation study is the work of many, whose contributions are combined to achieve the whole.

The chapters that follow will establish the skills and techniques required of construction professionals as participants in cost segregation studies. Chapter 7 starts by providing guidelines for reading and understanding plans and specifications.

Chapter 7
The Plans & Specifications

To accurately estimate the cost of a building asset, it is necessary to understand and be able to work with the language and symbols used in the working drawings, or construction plans. It is also important to be familiar with the technical specifications and their interdependency with the plans. This chapter provides an overview of these construction documents in terms of how they are organized and the information they provide for estimating.[1]

The Role of the Plans

The graphic portion of building project documents is commonly referred to as the *plans, drawings*, or *blueprints.* In this book, these terms are treated as synonymous and used interchangeably. Called by any of these names, the plans are the graphic representation or illustration of the project and comprise the lines, symbols, and abbreviations printed on paper that represent the owner's criteria, as interpreted by the architect, for the facility that is to be constructed. The plans, together with the specifications (discussed later in this chapter), make up the contract documents and form the basis of the contract for construction.

Design Development

Most drawings develop over several generations of review and modification as a result of owner input, coordination between the design disciplines, building code compliance, and general fine-tuning. This process is referred to as *design development* and occurs before the release of the final version of the drawings, called the *working drawings.* Working drawings are the completed design—a code-compliant representation of the project, ready for construction.

The completed drawings become a "set," which incorporates all adjustments, changes, and refinements made by the architect or engineer as the final step in design development. Working drawings should comply with all applicable building codes, including any local ordinances having jurisdiction. They should also include all of the information needed to identify, quantify, and estimate an asset. The set of working drawings consists of various disciplines of design, including the architectural or *core drawings,* structural engineering to ensure that the structure will support the imposed loads, and mechanical and electrical engineering to make the space habitable and functional.

All buildings are constructed with a definitive purpose and require professionals skilled in specific areas of design to make the overall facility design suitable. Just as most contractors develop their skills in one type of construction (commercial, residential, and so forth), design professionals focus on one general area of expertise. There are also specialty designers, such as for restaurant kitchens. Specialty drawings, included as part of the set, often require considerable coordination with the mechanical and electrical systems, as well as with the core drawings.

Other drawings in the set include designs that are less concerned with the structure itself than its support services. These are called *civil* or *site drawings.* They illustrate how support systems, such as the utilities, provide various services to and from the structure. *Grading and drainage plans* indicate how surface precipitation will be channeled away from the structure. Other drawings include *landscaping and irrigation design, paving,* and *curbing layout.* Ordinary site improvements, such as fencing, patios, walks, flagpoles, and the like, are shown on a kind of catch-all drawing called a *site improvements drawing.* Site improvements drawings are particularly important as they frequently show features that can be considered indicative to the trade or business of the building—helpful in determining features that are allowable for depreciation.

Some drawings are "cross-overs" in the sense that they show items of work or systems that may also be found in another set of drawings. For example, site drawings indicating lighting, power distribution, and low-voltage wiring (cable TV, telephone, and data) may also be shown on the electrical drawings.

Organization of the Working Drawings

There is a distinct organizational structure to the working drawings, which is almost universally accepted, and is as follows.

Architectural drawings (also known as core drawings) show the layout of the building and its use of space. Architectural drawings convey the structure's aesthetic value and show the dimensions and placement of all key features. The first architectural drawings in a set generally show large areas in less detail. As you progress through the architectural set, the level of detail increases. These drawings are prefixed by the letter "A" and are sequentially numbered.

Structural drawings illustrate the way in which the various load-carrying systems will transmit live and dead loads of the structure to the earth. Structural design is based on the architectural features, and relates to the core drawings. (For example, columns and beams are designed to avoid interrupting a space.) Structural components are prefixed by the letter "S" and are sequentially numbered.

Mechanical drawings show the structure's physical systems, such as plumbing, fire suppression/protection, and HVAC (heating, ventilating, and air conditioning) systems. These drawings may be prefixed by the letter "M" for mechanical, or "H" for heating. Plumbing drawings use the letter "P," and fire suppression drawings use "FP" (fire protection), "SP" (sprinkler system), or "F" (fire). The drawings are all sequentially numbered and shown in plan view.

Electrical drawings illustrate the electrical requirements of the project, including power distribution, lighting, and low-voltage specialty wiring, such as for fire alarms, telephone/data, and technology wiring. They often show the provision for power wiring of equipment illustrated on other types of drawings. They are prefixed by the letter "E" and are sequentially numbered.

Specialty drawings show the unique requirements of various spaces' special uses (such as kitchens, libraries, retail spaces, and home theaters). They define the coordination among other building systems, most commonly the mechanical and electrical systems. The drawings are sequentially numbered, and named by type. For example, "K" might be used for kitchen drawings, "F" for fixture drawings, and so forth.

Site drawings illustrate the structure's relationship to the property, including various engineering improvements to the site, such as the sanitary system, utilities, paving, curbing, and so forth. Site drawings are numbered sequentially, but have a less formal naming convention, open to the interpretation of the design engineer. They are easily distinguished from the core drawings, since they deal with the site only.

Drawings for each of the above categories will show only the work of their particular discipline. All lines and symbols that are not specifically related are shown in a lighter line weight, or "grayed-out." This helps coordinate or locate the work of a specific drawing with other drawings that indicate adjacent, but unrelated work.

Other Elements in the Drawings

In addition to the drawings just described, building plans also include a cover sheet, title block, and revisions. (Some revisions will be encountered on every drawing.)

The Cover Sheet

The cover sheet provides basic, but essential information, such as the name of the project; its location; and the names of the architects, engineers, owners, and consultants involved in the design. The cover sheet also lists the drawings that comprise the set in the order they will appear. The drawing list is organized by the number of each drawing and the title of the page on which it appears. The cover sheet may also list information specifically required by the building code having jurisdiction over the design of the project, including the total square foot area of the structure, the building code use group the structure will fall under, and the type of construction.

Another important element on the cover sheet is a list of abbreviations or graphic symbols used in the drawings. There is often a section that contains "general notes," such as, "All dimensions shall be verified in the field," or "All dimensions are to face of masonry." These notes help set standards for background information provided throughout the drawings.

Title Blocks

The title block is located in the lower right-hand corner of each drawing and should include the following information:

- The prefixed number of the sheet (so that it may be identifiable as to the group and order in which it belongs)
- The name of the drawing (e.g., "First Floor Plan")
- The date of the drawing
- The initials of the draftsperson
- Any revisions to the final set of drawings

The date and scope of the revisions should be noted within the title block. If there is not enough space available, the revisions should be noted close to it. The title block should specify whether the entire drawing is one scale, or whether the scale varies per detail, as in the case of a sheet of details. Sets of drawings for commercial projects require the stamp (and usually the signature) of the architect or engineer responsible for the design.

Graphic Formats (or Views) Used in Drawings

Architects and engineers use accepted standards or methods to present graphic information. Showing different views of the designed structure ensures that all required information appears on the drawings. There are six main graphic formats:

- Plan view
- Elevations
- Sections
- Details
- Schedules
- Diagrams

Each method illustrates various aspects of a project from a different viewpoint. The information is most effectively presented when multiple views are used together. Showing the same item in different views helps confirm and add to the information shown in a single view.

Plan View

The most common method, plan view, is presented as if you were looking down on the space. Plan views form the basis of the project, and often provide the most complete view. The most common plan view is the *architectural floor plan,* which shows doors, windows, walls, and partitions.

Variations of plan views include *structural, fire suppression, plumbing, HVAC,* and *electrical plans.* Each shows the work of the respective trades in plan view as they fit into the architectural floor plan. Other types of plan views include *reflected ceiling plans* and *partial plan views,* which illustrate a particular area and enlarge it for clarity. Partial views are most often used in areas of high congestion or detail. *Demolition plans* show proposed changes to the existing floor plan. *Roof plans* show the roof layout as it would be seen from overhead.

Plan views provide dimensions, which are necessary to calculate areas. Dimensions should be accurate, clear, and complete, showing both exterior and interior measurements of the space. The plan view provides the basis from which other drawings or views are further developed.

Elevations

Elevations provide a pictoral view of the walls of the structure, similar to a photograph of a wall taken perpendicular to both the vertical and horizontal planes. Exterior elevations may be titled based on their location with respect to the headings of a compass (north, south, east, or west elevation), or their physical location (front, rear, right side, or left side elevation.) The scale of the elevation should be noted either in the title block or under the title of the elevation.

Interior elevations provide views of the walls of the inside of a room. They illustrate architectural features, such as casework, standing and running trims, fixtures, doors, and windows. Exterior elevations provide a clear depiction of doors and windows, often using numbers or letters shown inside circles to show types that correspond to information provided in the *door and window schedule.* In addition, elevations show the surface materials of walls, and any changes within the plane of the elevation or façade. While the floor plan shows measurements in a horizontal plane, elevations provide measurements in a vertical plane with respect to a horizontal plane. These dimensions provide a vertical location of floor-to-floor heights, window sill or head heights, floor-to-plate heights, roof heights, ceiling heights, or a variety of dimensions from a fixed horizontal surface. You can use these measurements to calculate quantities of materials needed.

Building Sections

A building section, commonly referred to as a *section,* is a vertical slice or cut-through of a particular part of the building. It offers a view through a part of the structure not found on other drawings. Several different sections may be incorporated into building sections. Sections taken from a plan view are called *cross-sections.* Those taken from an elevation are referred to as *longitudinal sections,* or simply *wall sections.* Wall sections provide an exposed view of the building components and their arrangement within the wall itself. Referring to sections, in conjunction with floor plans and elevations, makes it possible to visualize the structure's composition.

Details

For greater clarification and understanding, certain areas of a floor plan, elevation, or a particular part of the drawing may need to be enlarged. This enlargement provides critical information about a building element that may not be available in other views. Enlargements are drawn to a larger scale and are referred to as *details.* Details can be found either on the sheet where they are first referenced, or grouped together on a separate detail sheet included in the various disciplines they reference. The detail is shown in larger scale to provide additional space for dimensions and notes. Details are not limited to architectural drawings, but can be used in structural and site plans and, to a lesser extent, in mechanical or electrical plans.

Schedules

In an effort to keep drawings from becoming cluttered with too much printed information or too many details, architects have devised a system to organize all types of repetitive information in an easy-to-read table, known as a *schedule*. Schedules list information pertaining to a similar group of items, such as doors, windows, room finishes, columns, trusses, and light or plumbing fixtures. The most common types are door, window, and room finish schedules, but information on any repetitive type of item can be assembled into a table and incorporated into a set of drawings.

Schedules are not limited to architectural drawings, but can be found in any discipline included within the set. A typical lighting schedule lists each light fixture by manufacturer and model, and provides information on size, voltage, and any related circuiting.

Diagrams

Some of the information presented in the set of drawings is more diagrammatical than pictoral. A *diagram* illustrates how the various components of a system are configured, and is often provided for purposes of coordination. Diagrams are commonly used for mechanical and electrical drawings, because of the complex nature of the work. Diagrams often illustrate work that has a vertical configuration. Common examples include diagrams for fire alarm risers, waste and vent piping risers, and fire protection.

Drawing Conventions

Certain conventions have been adopted to provide consistent drawings from one design firm to another. The most common graphic features are lines, in-fill techniques, and shading, which can often contain subtle, but important information related to the detail shown. While most of these conventions are widely accepted and practiced, there will always be minor deviations based on local practices. This is most apparent in the use of abbreviations and symbols, which can vary based on local practice or construction techniques.

Lines

Drawings must convey a great deal of information in a relatively small space, where there is no room for a lot of wording. Consequently, different types of lines are used to communicate information. The most common ones are discussed below.

Main object line: A thick, heavy, unbroken line that defines the outline of the structure or object. Used for the main outlines of walls, floors, elevations, details, or sections.

Dimension line: A light, fine line with arrowheads or "tic" marks at each end, used to show the measurements of the main object lines. The arrowheads fall between extension lines that extend from the main object lines to show the limits of the item drawn. The number that appears within the break in the dimension line is the required measurement between extension lines.

Extension line: A light line that extends from the edge or end of the main object line. Used together with dimension lines to help determine the quantity of a particular feature.

Hidden or invisible line: A light, dashed line of equal segments that indicates the outlines of an object hidden from view, under or behind some other part of the structure, such as a foundation shown in elevation that would be below grade.

Center line: A light line of alternating long and short segments that indicates the center of a particular object. Frequently labeled with the the letter "C" superimposed over the letter "L."

Material Indication Symbols and Shading

In-filling certain graphic features on a drawing helps convey their content or composition. In-fills are called *material indication symbols* and can be used to indicate whether the feature is solid, as in the case of cast-in-place concrete, or hollow, as with concrete masonry block. Because of the different views used on drawings, various materials must be recognizable at each view, from plan to section to elevation. As with abbreviations, material indication symbols are subject to change based on specific materials used in various parts of the country.

Shading

Architects and engineers can convey information in a more subtle manner by changing the intensity of a particular feature. This effect, called *shading,* increases or decreases the focus on the item, merely by its intensity. Items in the foreground or focus are often drawn darker or thicker. Objects in the background are lighter in color, and drawn less sharply. Shading is often used to differentiate between proposed and existing work on renovation projects.

Graphic Symbols

Graphic symbols are another standardized way to show repetitive information on drawings. *Section markers* indicate where a section is cut through an object, and can be directional or non-directional. *Elevation symbols* direct you to the drawing that contains a noted elevation. They indicate differences in vertical height, such as the distance between floors, and provide a reference point to use in calculating the height of components in walls or partitions.

Design professionals often draw a feature, and to save space on the page, use a *break in a continuous line.* This symbol conveys that the feature is not drawn to scale. *Geometric shapes* with letters, numbers, or dimensions within the shape define certain features or main objects. This graphic symbol is frequently used to name windows, doors, rooms, partition types, plumbing or electrical fixtures, and ceiling heights. The number or letter within the symbol is often tied to a schedule.

Trade-Specific Symbols

Like graphic symbols, *trade-specific symbols* depict items that are common to the various trades. Because of the highly diagrammatic nature of mechanical and electrical drawings, there is an abundance of unique, trade-specific symbols in use. Engineers typically provide legends that define the various symbols. Some, such as for a water closet or toilet, are highly recognizable because they mirror the feature in real life.

Abbreviations

Abbreviations are used to save design professionals time, as well as space, on drawings. Many different abbreviations are used in daily practice, but it is not necessary to memorize them all. It is standard practice to list them on the cover sheet of the set of drawings.

Scale

Since there are obvious physical limitations to drawing a building's actual size on a piece of paper, the drawings retain their relationship to the actual size of the building using a ratio, or scale, between full size and what is seen on the drawings. There are two major types of scales: the *architect's scale* and the *engineer's scale.*

Architect's Scale

The architect's scale is used for building drawings, as well as the engineering disciplines. The actual architect's scale may be flat, like a ruler, or three-sided. The three-sided architect's scale has 10 separate scales: 1/8" and 1/4", 1" and 1/2", 3/4" and 3/8", 3/16" and 3/32", and 1-1/12" and 3". The one remaining side is in inches, similar to a ruler. For example, when used on a floor plan that is 1/4" scale, each 1/4" delineation represents one foot. The same rules apply for 1/8" scale, in that each 1/8" segment on the drawing represents 1'-0" of actual size. The same approach applies to each of the other scales. There is no strict convention that states which scale should be used on which drawings. In general, as the area of detail being drawn becomes smaller, the scale often increases. For example, a floor plan may be fine at 1/4" = 1'-0", yet the detail of an element within that floor plan would be better illustrated in 1/2" or 3/4" = 1'-0" for clarity.

Engineer's Scale

The engineer's scale is similar to the architect's scale and is typically (though not exclusively) used to prepare civil drawings. It is used to indicate distances on site plans, when these distances are greater than would be encountered in the plans for the building. The difference is the size of the increments on the sides of the scale. The engineer's scale has six scales: 10, 20, 30, 40, 50, and 60. For example, the 10 scale refers to 10' per inch; the 20 scale is 20' per inch, and so on. Other specialty scales are divided into even smaller increments, such as 100.

Civil Drawings

Commercial projects such as shopping malls, restaurants, and office buildings, to name a few, include a *site plan,* which illustrates the relationship of the proposed structure to the building's lot, as well as the various site improvements needed to accommodate the new building. The grouping of different types of site drawings, such as utility and drainage, grading, site improvement, and landscaping plans, are known under the general classification of *civil drawings.* Civil drawings encompass all work other than the structure itself. They have some unique conventions and nomenclature that merit a separate review. The most obvious difference between civil drawings and architectural drawings is the use of the engineer's scale. As mentioned earlier, smaller scales are used on site drawings to indicate much larger areas. It is important to note the scale in order to avoid errors in measuring during the takeoff. To avoid confusion, it is best to use the title block to clarify the type of drawing and scale.

The following is a review of the more common terms and symbols associated with the various civil drawings.

Site Plan

The main purpose of the site plan is to locate the structure within the confines of the building lot. Even the most basic site plans clearly establish the building's dimensions, usually by the foundation's size and the distance to property lines. The latter, called the *setback dimensions,* are shown in feet and hundredths of a foot, versus feet and inches on architectural drawings. For example, the architectural dimension of 22'-6" would be shown as 22.50' on a site plan. This decimal system is used because it is the basis of measurement for the land surveyor, who is the predominant engineer used to lay out the site.

As a starting point for the site design, a site survey is performed by a registered land surveyor, who also records special conditions. These may include existing natural features, such as trees or water, as well as man-made improvements, such as walks, paving, fences, or other structures. The new site plan shows how the existing features will be maintained, modified, or removed to accommodate the new design.

Another chief purpose of the site plan is to show the unique surface conditions, or *topography,* of the lot. Changes in the elevation of the lot, such as slopes, hills, valleys, and other variations in the surface, are shown on a site plan by means of a *contour,* which is a line connecting points of equal elevation. An *elevation* is a distance above or below a known point of reference, called a *datum.* The datum could be sea level, or an arbitrary plane of reference established for the particular building. For projects in which the topography must be shown separately for clarity, a *grading plan* is used.

A known elevation on the site for use as a reference point during construction is called a *benchmark.* The benchmark is established in reference to the datum, and is commonly noted on the site drawing with a physical description and its elevation relative to the datum. For example: *"Northeast corner of catch basin rim - Elev. 102.34"* might be a benchmark found on a site plan. When individual elevations, or spot grades, are required for other site features, they are noted with a "+" and then the grade. For example: *"+123.45"* would designate a spot grade for a particular feature. (Grades are accurate to two decimal places, whereas contours are expressed as whole numbers.) Some site plans include a small map, called a *locus,* showing the general location of the property in respect to local highways, roads, and adjacent pieces of property.

Drainage and Utility Plans

Larger projects will have several site plans showing different scopes of related or similar work, such as drainage and utility plans. Utility drawings show locations of water, gas, sanitary sewer, and electric utilities that will service the building. Drainage plans detail how surface water will be collected, channeled, and dispersed on or off site. Both plans illustrate, in plan view, the size, length, and type of pipes and special connections or terminations of the various piping. Because the effluent in certain types of pipe moves by gravity, the elevation of each end of the pipe must be different.

Certain site plans require clarification in the form of a detail, similar to the architectural detail. Classic examples are sections through paving, pre-cast structures, pipe trenches, and curbing. Details are not limited to scaled drafting, but occasionally appear in the form of perspective drawings, which are not drawn to scale and are used as a means of clarification only.

Landscaping Plans

Landscaping plans show the location of various species of plantings, as well as lawns and garden areas. The plantings are noted with an abbreviation, typically three letters, along with the quantity of the particular species. This designation corresponds to a planting schedule, which is a complete listing of plantings by common name, Latin or species name, and quantity and size. Notes describing planting procedures or handling specifications accompany the schedule. Irrigation drawings may be included to illustrate how the landscaping elements should be watered.

Paving/Curbing Layout Drawings

To accurately show the layout of parking lots and driveways, a *paving/curbing layout drawing* is needed. This plan shows the various types of bituminous, concrete, and brick paving and curbing, and the limits of each. Again, it is important to review the legend symbols in order to clearly delineate where one material ends and another begins. Details showing sections through the surface are used to differentiate between its thickness and the substrate below.

Site Improvement Drawings

When the project warrants, separate drawings may be needed to clarify various site improvements, such as walks, retaining walls, patio paving, fences, steps, benches, play areas, and flagpoles. As mentioned earlier, site improvement drawings are often used as a "catch-all" for miscellaneous items that do not fall neatly into one of the other classifications of work.

Existing Conditions Site Drawings

For projects with existing drainage, utilities, and structures, an *existing conditions plan* is invaluable for understanding and calculating the difference between actual conditions and proposed work. The existing conditions are shown in the background "grayed out" or lightly shaded, and the new work is shown darker in the foreground. Sometimes test boring logs are provided; these document engineering tests conducted to determine the load-bearing and general quality of the subsurface at the site.

As-Built Drawings

Once the project has been completed, project owners, architects, and even local building officials often require that a set of record drawings, called *as-built drawings,* be produced. As-built drawings record for posterity any subtle changes between the working drawing design and the actual way the structure is built. For a variety of reasons, it is not uncommon for the contractor, with the approval of the architect or engineer, to make minor changes in the locations of certain features, such as piping (either underground or within plenum space), or dimensions that have no decernable effect on the building, but are necessary for coordination of the work. Other changes or modifications initiated by the owner or architect, in the form of *change orders,* can be of a more dramatic nature. These changes can involve virtually any aspect of the project, and are also recorded on the as-built drawings. As-built drawings are typically prepared for the entire set of working drawings that were used to construct the building. A well-prepared set of working drawings can be very useful as a graphic "snapshot" in locating, defining, and quantifying assets for recovery.

The Specifications

Architects and engineers use *technical specifications* to establish the quality of the materials and workmanship to be incorporated in the project. The specifications identify an acceptable level of quality that becomes the basis for the contract. The specifications are intended to be used in conjunction with the drawings. If the drawings are the *quantitative* representation of the project shown in a graphic format, then the "specs" (as they are commonly referred to) are the *qualitative* requirements of the project described in a written document.

The Role of the Specifications

The specifications are part of the contract documents, along with the working set of drawings, and define in detail the processes and materials for the project. Technical information about the quality of materials and workmanship is not always incorporated on the drawings themselves, due to lack of space and the need to maintain clarity. For commercial construction projects, working drawings are issued with a separate set of specifications.

The specs perform a variety of functions, including:

- Serving as the legal basis for the Contract for Construction.
- Defining the quality or grade of materials to be used in the project.
- Defining the acceptable workmanship or providing standards to judge workmanship.
- Providing guidelines for resolving disputes between parties to the contract.
- Providing a basis for accurately estimating cost.
- Complementing the graphic portion of the project, the drawings.

Technologies, processes, and products are continually evolving in the construction industry, and architects and engineers are incorporating these advancements into their designs. As a result, highly specific information is needed in the specifications. The materials and processes are described in such detail that the intent of the designer, as well as the product or system, can be upheld in case of a dispute, or if products are installed incorrectly.

Organizing Specifications by CSI MasterFormat

The most widely accepted system for arranging construction specifications and estimates is called CSI MasterFormat. Developed by the Construction Specifications Institute, the MasterFormat system is used for organizing the technical specifications for construction products and services. CSI has allocated a five-digit code and topic descriptions to all components of the specifications. MasterFormat organizes the technical specifications information into 16 divisions, each of which is a grouping of similar or related work numerically organized into subsections.

The 16 CSI MasterFormat divisions were determined based on relationships of activities in the actual construction process, and roughly follow the natural order of the construction of a building. The specification divisions and a general summary of their contents are as follows:

Division 1 – General Requirements: A summary of the work, as well as the definitions and standards for the project and project coordination, meetings, schedules, reports, testing, samples, submittals, shop drawings, close-out, cleanup, quality control, and temporary facilities. In addition, Division 1 addresses pricing issues, such as unit prices, alternates, and allowances.

Division 2 – Site Construction: Clearing of the site, earthwork, site drainage, utilities, roads, walks, paving, curbing, general site improvements, subsurface investigations, landscaping, and heavy site work, such as shoring, pile driving, and caissons.

Division 3 – Concrete: Formwork, reinforcing, pre-cast and cast-in-place concrete, and cementitious decks.

Division 4 – Masonry: Brick, block, stone, mortar, anchors, reinforcement, and masonry restoration and cleaning.

Division 5 – Metals: Structural steel, metal joists, metal decking, light-gauge framing, and ornamental and miscellaneous metals.

Division 6 – Wood and Plastics: Rough and finish carpentry, millwork, casework, and plastic fabrications.

Division 7 – Thermal and Moisture Protection: Waterproofing, dampproofing, insulation, roofing, siding, caulking, and sealants.

Division 8 – Doors and Windows: Metal and wood doors and frames, windows, glass and glazing, and finish hardware.

Division 9 – Finishes: Gypsum wallboard systems, board and plaster systems, painting and wallcoverings, flooring, carpeting, acoustical ceiling systems, and ceramic and quarry tile.

Division 10 – Specialties: Items such as demountable partitions, toilet partitions and accessories, fire extinguishers, postal specialties, flagpoles, lockers, signage, and retractable partitions.

Division 11 – Equipment: Specialized equipment for restaurants, retail stores, banks, gymnasiums, schools, churches, laboratories, prisons, libraries, and hospitals.

Division 12 – Furnishings: Cabinetry, rugs, tables, seating, artwork, and window treatments.

Division 13 – Special Construction: Greenhouses, clean rooms, swimming pools, integrated ceilings, incinerators, and sound and vibration controls.

Division 14 – Conveying Systems: Elevators, escalators, lifts, dumbwaiters, cranes, chairlifts, and hoists.

Division 15 – Mechanical: Fire protection systems, plumbing, heating, air-conditioning, ventilating, gas piping, special piping, refrigeration, and controls.

Division 16 – Electrical: Electrical service and distribution, wiring devices, fixtures, communications, and power.

The group of tasks included within each of the 16 divisions of CSI MasterFormat is numerically organized with a five-digit code. The first two numbers of the code identify the division. The next three digits further define the type of work within the division.

Preparation of the project manual that contains the technical specifications is a substantial task, primarily the responsibility of the architect. Individual disciplines, such as mechanical, electrical, and structural engineers, review, edit, and contribute to their individual sections of the technical specifications. The technical writer must create a complete document using very specific language that will guide the contractor in the bidding and building processes, and will also serve as a powerful tool to enforce the contract. The project manual calls for a skillful use of language, a high level of proficiency in understanding and coordinating technical information, and the ability to process that data into usable information.

Technical Specifications

The technical specifications define the scope, products, and execution of the work. This is the "meat and potatoes" section, providing the practitioner with the necessary information (in a highly organized and industry-accepted format) to accurately price assets or the entire structure. The technical sections provide the following information for each activity:

- Administrative requirements
- Quality or governing industry standards
- Products and accessories
- Installation or application procedures
- Workmanship requirements

The three-part format of the technical sections provides a consistent organizational system for locating pertinent information quickly and efficiently:

- Part 1 – General
- Part 2 – Products
- Part 3 – Execution

Part 1, General

Part 1, the general section of the specifications, provides a summary of the work included within that particular section. It ties the technical section to the General Conditions and Supplementary General Conditions of the Contract, which is an essential feature in maintaining continuity between the general contractor and subcontractors. Part 1 identifies the applicable agencies or organizations by which quality assurance will be measured. It defines the scope of work that will be governed by this technical section, including, but not limited to, items to be furnished by this section only, or furnished by others and installed under this section. It also identifies other technical sections that have potential coordination requirements with this section, and defines the required submittals or shop drawings for the scope of work described in this section. Part 1 also establishes critical procedures for the care, handling, and protection of work within this section that may have an additional cost impact.

Part 2, Products

Part 2 deals exclusively with the products and materials to be incorporated within this technical section of the work. For products that are directly purchased by the contractor from a manufacturer or supplier, the items can be identified using one of three methods:

- Proprietary specifications
- Performance specifications
- Descriptive specifications

Proprietary Specifications These specifications spell out a product by name and model number. Proprietary specifications have the unique advantage of allowing the architect or owner to select a product they desire or have used successfully on prior projects. The advantage of requiring specific products is the level of reliability they provide. The disadvantage is that they eliminate open competition. To help reduce the exclusivity of the proprietary spec, the architect often adds phrasing called the "or equal" clause, which allows limited competition.

Performance Specifications An alternate method of specifying products and materials is based less on makes and models and more on the ability to satisfy a design requirement or perform a specific function. This type of specifying is called a *performance specification.* In lieu of specifying a particular product by name, the architect opens competition to all products that can perform the specific functions required to complete the design. This approach allows healthy competition among various manufacturers that have a similar line of products. It ensures competitive pricing and more aggressive delivery schedules.

Descriptive Specifications The last method of specifying a product or process is by using *descriptive specifications,* which are written instructions or details for assembling various components to comprise a system or assembly. Most often, descriptive specifying is used for generic products such as mortar, lumber, or concrete. Frequently, no manufacturers' or proprietary names are mentioned or needed.

Part 3, Execution

Part 3, called the *Execution,* deals exclusively with the method, techniques, and quality of the workmanship. This section makes clear the allowable tolerances of the workmanship. The term "tolerances" refers to plumb, straight, level, or true. The Execution section should also describe any required preparation to the existing surfaces in order to accommodate the new work, as well as a particular technique or method for executing the work. This method or technique should be taken into account by the practitioner when deriving the quantities of the asset. In addition, the execution portion of the specification section defines workmanship quality for pricing purposes. Terms such as "first-class workmanship" often indicate that the cost of labor is higher than what would be defined as "ordinary."

Part 3 of the specifications also addresses issues such as fine-tuning or adjustments to the work after initial installation, general cleanup of the debris generated, final cleaning, and protection of the work once it is in place. Some sections of Part 3 may identify any ancillary equipment or special tools required to perform the work, such as staging or scaffolding. These are also part of the assets' costs.

Modifications to the Contract Documents

Addenda

The bidding process often produces questions that require answers or clarification from the architect or engineer. Any changes to the contract documents made during the bidding period (the time period beginning on the date the drawings are issued and ending on bid day) in the form of modifications, clarifications, or revisions, for any reason, are called *addenda.* An addendum can be issued only by the architect—in writing—and will automatically become part of the contract documents, complete with all of the rights and benefits of the Contract for Construction and the General Conditions of the Contract. Any addenda issued should be reviewed, along with the plans and specifications, for the changes that may affect specific assets.

Alternates

Often, owners want to see how a change in materials, method of construction, or addition or subtraction of work will affect the project's price during the bid phase of the project. These requests are presented in the form of additions or deletions to the base price, called *alternates.* Typically, the alternate is listed at the end of the specification section that is affected by it, and also in Division 1 under the section, "Alternates." Be sure to review any alternates that were selected during the contract award for their cost impact to the project. In many projects, alternates involve decorative elements and are priced separately so that if there are "sufficient funds" in the budget, the alternate can be added. An example is the additional cost for raised panel woodwork wainscoting in a boardroom. This clearly could be a significant item in a cost segregation study. The cost of that work should have been identified separately, prior to the commencement of the study.

Allowances

Occasionally, at the time the contract documents are ready to be issued, some items have not yet been finalized and are not ready for inclusion in the bid set. Rather than leaving the item out altogether, the designer includes a cash allowance. The allowance is a fixed lump sum, such as "$10,000 for the furnishing and installation of exterior signage lighting." The allowance can also be in the form of a unit price, such as, "an allowance of $45.00 per square yard (SY) for carpet, including installation and delivery charges."

Typically, it is clearly stated just what the allowance is for: materials, furnished and delivered only, materials and labor, or the entire scope of work. At the completion of the project, the actual cost is computed for items included as allowances, savings are returned to the owner, and overages are added to the contract price. This "adjustment" frequently occurs as a contract modification or change order that can be tied to the allowance. Change orders would be evident in contract billings. *(See Chapter 11, "Sources of Pricing Information.")*

Unit Price

In designing some projects, architects or engineers are sometimes unable to provide sufficient detail in the drawings so that the estimator can determine an exact quantity of a certain task or activity. A good example would be computer connection ports. The architect or engineer may be aware of the work to be done and the techniques or quality required, but is unable to determine the exact amount of connection ports required. In an effort to at least establish the cost of this work for contract purposes, unit prices are requested and submitted as part of the bid form or proposal. The unit prices, if accepted, become part of the contract, and payment is based on the quantity installed under the contract multiplied by the unit price to arrive at the compensation due. Again, when complete, the unit price work is issued as a contract change order that can be directly reconciled back to a unit price.

Conclusion

Once the plans and specifications have been studied, addenda or contract modifications have been reviewed, and a site visit has been conducted, then the cost segregation practitioner can begin the quantity survey or takeoff of the assets that qualify for recovery. See Chapter 8 for detailed discussion of takeoff and estimating procedures.

[1] In depth coverage is available in *Plan Reading & Material* Takeoff and *Estimating Building Costs,* published by RSMeans.

Chapter 8

Estimating Standards & Practices

In the construction industry, there are standards and accepted practices for establishing material and labor quantities and estimating their costs. Most of these methods have developed over time as a result of what works best and is most accurate. Accuracy is essential in allocating asset costs for a cost segregation study, and using accepted standards adds to the professionalism of the study.

The estimating process can be broken down into two phases: quantity survey (or "takeoff") and pricing. This chapter will focus on the former. Pricing will be covered in Chapter 11.

The Contract Documents

If contract documents for a facility are available, they should be thoroughly reviewed before the value of assets is estimated. The contract documents most often consist of:

- A set of plans
- A set of specifications
- Applications for payment or other types of invoicing
- Change orders or contract modifications
- Any relevant addenda or bulletins
- Additional relevant documents, such as geotechnical reports and documentation, or unique requirements of local authorities. (Examples include conservation or historic commission directives or conditions and/or amendments added to the documents by local inspection officials during the building code review.)

The plans and specifications contain equally important, but distinctly different, information. The information illustrated on the plans should be supported by the written language of the specifications. Neither is intended to stand alone, and both are used, along with other documents noted above, to prepare a complete estimate. Reviewing the documents is essential to identify references to assets. During this review, estimators usually make notes of questions or needed clarifications. Often, answers are uncovered as the review proceeds.

When Contract Documents Are Not Available

When there are no plans and specifications available, as is often the case in cost segregation studies on buildings acquired or constructed in the past, quantities for asset estimating are obtained through a site visit. Physical measurements of site improvements, building square footage, structural components, and personal property assets all need to be field-verified, along with asset quality and condition. Site and building checklists and/or forms are often utilized to record the field data.

The Quantity Takeoff Process

After an initial document review and when all questions have been answered, the quantity takeoff begins. Formally called the "quantity survey," the takeoff breaks the project down into its incremental parts, called *tasks* or *activities* (terms used interchangeably) to refer to actual units of work to be performed. Examples of different tasks might include: *"place and finish concrete slab," "form footings," "frame exterior walls,"* or *"paint interior doors and frames."*

Tasks are quantified by details and dimensions provided on the drawings. They are also *qualified* by the characteristics of their materials and labor as required by the specifications. These two components, quantity and quality, are derived from the plans and specifications, respectively, and establishing them is an essential part of accurate estimating.

Tasks or activities are made up of the various components of material, self-performed labor (performed by the general contractor's own employees), tools and/or equipment, subcontractors, and, occasionally, non-production-related expenses. Not every task involves every one of these components, and some have a combination. For example, *"paint interior doors and frames"* has both a material and labor component only if the work is self-performed. If the work is performed by a subcontractor, the takeoff might include only the subcontractor component. If the work is self-performed using spray equipment, the task includes material, labor, and equipment components.

Guidelines for an Accurate Takeoff

A precise and thorough takeoff is the basis of a sound estimate. Errors in the quantity takeoff portion of the estimate can be compounded during the pricing phase, regardless of how precise the unit price may be. The following guidelines are based on common sense practices and should help prevent, or at least minimize, errors while making the takeoff process faster and more efficient.

1. Write Clear Task Descriptions

The takeoff description should be clear and legible and should indicate the work needed and the part of the structure involved, as well as the location on the drawing where the quantity originates. An example of a thorough task description would be:

Product display lighting, fixture Type A in the showroom.

2. Use Industry-Accepted Units

A takeoff is a descriptive list of activities with quantities derived from the dimensions on the documents. These quantities are extended into accepted units for pricing. For example, concrete is estimated in cubic yards (CY) because it is sold by the cubic yard, the industry standard.

3. Follow a Logical Order

The takeoff should follow a logical, organized pattern. The best approach is to proceed in the same order as the structure is built—from the ground up, using the Construction Specifications Institute (CSI) MasterFormat divisions as a guide. Each asset should be taken off and organized by division. For example, all the work in Division 3, Concrete, should be complete before moving on to an asset in Division 4, Masonry.

4. Review Scales, Notes, Abbreviations, and Definitions

Review drawings and details carefully for notes and scale. Remember that the scale can change from one drawing to another. A general rule of thumb is that as the detail becomes smaller in scope, the scale becomes larger. Whenever possible, use the printed dimensions shown on the drawings. Scaled dimensions should be used only when printed dimensions are not available.

5. Be Consistent

Develop a systematic approach when working with drawings. For instance, when taking measurements, move in a clockwise direction around a floor plan. Begin counting similar features, such as light fixtures, from left to right and top to bottom. Whatever the procedure, use a standard, consistent approach.

6. Number Takeoff Sheets

Number each takeoff sheet and keep them in order. Whenever possible, similar tasks that relate to the same asset should be kept together.

7. Use Decimals

Decimals are preferable to fractions in the quantity takeoff because they are more precise and easier to use with a calculator. Dimensions on drawings should be converted to their decimal equivalent.

8. Verify Appropriate Level of Accuracy

While accuracy is important, over-accuracy is most often a waste of time. There is an old adage in the construction business that warns, *"Don't spend $10 of estimating time figuring a $1 item . . . unless there are hundreds of them."* The level of accuracy should be commensurate with the task being taken off or estimated. Rarely is it necessary for a number to be calculated to more than two places after the decimal.

Remember, quantity takeoff is performed for cost estimating purposes. Since estimating is not an exact science, the quantites need not be overly precise. Reasonable accuracy is necessary and should be expected. Most items can be rounded to the nearest full unit.

9. Account for Waste

Some materials require an added allowance for waste. Waste is applicable to materials only and should not be applied to labor or confused with productivity. Before waste is added, the quantities are referred to as *net quantities.* After an allowance for waste has been added, quantities are referred to as *gross quantities.*

10. Double-Check the Takeoff

The quantity takeoff should be checked by someone other than the person who prepared it. Ideally, another complete takeoff and estimate would be created as a means of checking the first. However, this may not be practical or cost-effective. At the very least, random checks of specific quantities should be performed by another individual.

11. Mark up the Drawings as Work Papers

Items on drawings can be checked off, colored in, or highlighted with colored pencils or markers to indicate which items have already been taken off. Drawings are the work papers of the cost segregation study and should be kept as a record, providing the basis for how the quantities were derived. The most obvious benefit of retaining the drawings is substantiation of an asset during an IRS audit. Being able to demonstrate how the asset was quantified lends credibility to the cost segregation study.

12. Focus on the Task at Hand

Both the takeoff and pricing process require a high level of concentration. Frequent distractions or attempts to multi-task are often the greatest source of error. When the takeoff must be interrupted, select a natural stopping point and note it clearly.

13. Organize the Documentation

Careful organization and neatness of work papers and takeoff sheets are crucial. If supporting documentation is needed (including sketches or details as to how unusual features were estimated), it should be retained and attached to the pertinent quantity sheet. Even if the takeoff is performed electronically, there will still be work papers and notes. Pre-printed takeoff sheets and forms should be used whenever possible for written notes. All calculations should follow a logical, sequential process.

Waste—Net and Gross Quantities

Quantities derived during the takeoff process often do not reflect exactly what is actually purchased when the work is in progress. For example, the person who performs the takeoff may determine that the floor area to be covered with carpet is 3,000 SF. However, carpet is priced by the square yard (SY) and sold in rolls of specific widths. The difference between the quantity that must be included in the estimate and the quantity that is actually to be installed is called *waste.* Waste is applicable to materials only, and not every task has waste. Pricing is done at the gross quantity level, after the waste has been added—not on the net quantity. *(See Chapter 9 for more on calculating waste.)*

Price breaks based on total quantities should also be taken into account. Economy of scale (large-quantity material purchases) generally produces lower unit prices. This factor should be considered in reconciling asset costs.

Classification of Costs: Types of Costs

There are two main types of costs associated with the unit price estimate:

- Production costs
- Non-production costs

Production Costs

Production costs are part of the actual, physical project, including materials, labor, tools, equipment, and subcontractor costs for each task or activity incorporated in the final structure. If we analyze the unit price of an individual task, such as a concrete slab that is placed by a contractor's own employees, the individual components are: the concrete material itself, the labor to place and finish the concrete, and the equipment rental/ownership costs for power trowels. If the slab is placed and finished by a subcontractor, the costs might be estimated as a single cost (lump sum), which already includes each of the above components.

Non-Production Costs

Non-production costs are *overhead costs,* or the cost of supporting the production activities. Overhead costs are divided into two classifications:

- Direct overhead
- Indirect overhead

Both direct and indirect costs can be proportionally allocated to a particular asset, though they are inherently different. In some circumstances, such as use of a crane, 100% of a direct overhead cost might be allocated to a particular task.

Direct Overhead Costs Direct overhead costs, sometimes referred to as *project overhead,* are directly related to one project. These include the costs for items like temporary facilities, trailer rental, superintendence, telephone useage, staging or lifts, and electrical power. Direct overhead costs can be further segregated for estimating as:

- Time-sensitive costs
- Fixed costs

Time-Sensitive Costs These cost items are defined by the duration for which they are needed. The longer the item is on the job site, the higher the cost. For example, most contractors rent trailers with a monthly rental fee. The longer the project goes on, the higher the accrued cost for the trailer rental. Many other time-sensitive costs can be found in project overhead. To accurately assign a dollar value to them, the estimator (in the contractor's bid phase) develops a schedule to determine these costs over the term of the project.

Fixed Overhead Costs Direct overhead costs that are not time-related are classified as *fixed* project overhead costs. Examples include building permit fees, registered site layout, engineering design fees, and access roads or ramps. Usually there is a single occurrence for each of these costs.

Indirect Overhead Costs These costs are referred to as *main office overhead*—any general costs that are not unique to a specific project. Indirect overhead costs are the costs associated with being in business. Examples include main office rent or mortgage payments, salaries of officers and clerical staff, base insurance policies, company vehicle costs, and so forth. Indirect overhead costs are accumulated with time and as work is performed. An incremental portion of the indirect overhead is assigned to each project task, usually as a percentage.

The Value of Unit Price Estimates

A properly assembled unit price (or cost per unit of work) estimate is highly defensible, as it establishes detailed, accurate costs for individual assets. A unit price estimate should have a quantity and a price for each item of work or expense identified in the construction documents. A unit price estimate is performed by breaking a project down into tasks, quantifying those tasks, and then applying a price based on the number of units for each task. As mentioned previously in the Quantity Takeoff discussion in this chapter, each task can have various and multiple components, including materials, labor, tools, equipment, and subcontractors that make up the unit price. Following is a sample unit price (that includes the above elements) as it might appear in an estimate.

Task Description	Quantity	Unit		Unit Price		Total
Place and finish concrete slab	10,000	SF	@	$1.00	=	$10,000

The unit price of $1.00 per SF covers the labor and equipment used to place and finish the concrete slab. The $1.00 per unit cost is based on a unit of one SF. Once the unit price has been multiplied by the quantity, the resultant cost is called the *extended cost,* or total cost for the item.

Conclusion

Depending on the property, available documentation, and the level of investment in the cost segregation study, a cost segregation practitioner may not have the time or the need to follow all of the detailed estimating practices outlined in this chapter. However, every quantity and price should be able to be substantiated in factual terms in order for the estimate to be considered a professional product that will stand up to an audit by the IRS.

Chapter 9
Area & Volume Calculations for Takeoff

Calculating Linear Measure, Area, & Volume

To perform even basic quantity surveys or takeoffs, both construction cost estimators and cost segregation practitioners should be able to calculate linear measurements, surface area, and volumes of solids.

- Linear measurement refers to the measurement of lines in a single dimension.
- Area refers to the measurement of surfaces in two dimensions—length by width.
- Volume refers to the measurement of solids in three dimensions—length, width, and height or thickness.

This section will review the basic formulas and relationships needed for these calculations. The formulas are fairly simple and need not be memorized, but it is important to know which formula to use when, and where to find it.

Units of Measure

The most fundamental rule is to use the correct units of measure for area and volume. Area is expressed in *square* units, most often square feet (SF) or square yards (SY). Volume is expressed in *cubic* units, the most common being cubic feet (CF) or cubic yards (CY). Keeping the units the same is essential for accurate calculations. For example, feet multiplied by feet results in square feet, yards multiplied by yards results in square yards, and so on. Multiplying a dimension in feet by a dimension in inches leads to an erroneous value.

Decimal Equivalents

Estimators must convert all dimensions to the same units to yield a usable result. Often, a dimension on the drawings is given in feet *and* inches. For example, to calculate the area of a space that is 24'-6" x 20'-3", the dimensions must be converted to their decimal equivalents. The inch dimensions should be changed to feet to calculate total square footage of the area. Once this has been done, the values can be easily entered into a calculator. Following are the decimal equivalents of the 12 inches in a foot:

1" =	0.08 feet	7"	= 0.58 feet
2" =	0.17 feet	8"	= 0.67 feet
3" =	0.25 feet	9"	= 0.75 feet
4" =	0.33 feet	10"	= 0.83 feet
5" =	0.42 feet	11"	= 0.92 feet
6" =	0.50 feet	12"	= 1.00 feet

Note that the decimal equivalent of 2" in the table above is not precisely two times the decimal equivalent of 1". This is because two-place accuracy after the decimal point is sufficient for estimating purposes in most cases. (While the actual value of 2" in decimal form is .0166666 of a foot, the value is rounded up to 0.17.) In estimating, the decimal value is limited to two places after the decimal, and it is always rounded up. For example, if the total quantity of concrete in a takeoff is 34.22 CY, round to 35 CY for use in the cost analysis portion of the estimate.

According to the table above, it should hold true that if *1" = 0.08 feet, then 1/8" = 0.01 feet.* Using this simple mathematical analogy, it is easy to covert any dimension with accuracy to 1/8". For example:

To convert a dimension of 20'-3-5/8" to its decimal equivalent, start by converting the 3" to 0.25', then the 5/8" to 0.05'. Added together (0.25' + 0.05'), the result is 0.30'. This can be added to the 20', totaling 20.30'.

To calculate the area of a room that is 24'-6" x 20'-3 5/8", the decimal equivalent would be 20.5' x 20.30', which equals 497.35 SF.

Linear Measurement

Perimeter

Perimeter measurements apply to surfaces, such as floors and walls, or linear items, such as baseboard or trim. The side dimensions are added to produce a linear total, most often in feet (LF).

Angles & Triangles

If a rectangle with four 90° corners were divided in half by a line connecting the opposite corners, the resulting shape would be a *right triangle.* A triangle has three angles that total 180°. It also has three sides: base, altitude, and *hypotenuse.* (The hypotenuse is the diagonal line connecting opposite corners of the triangle, opposite the 90° angle in a right triangle.)

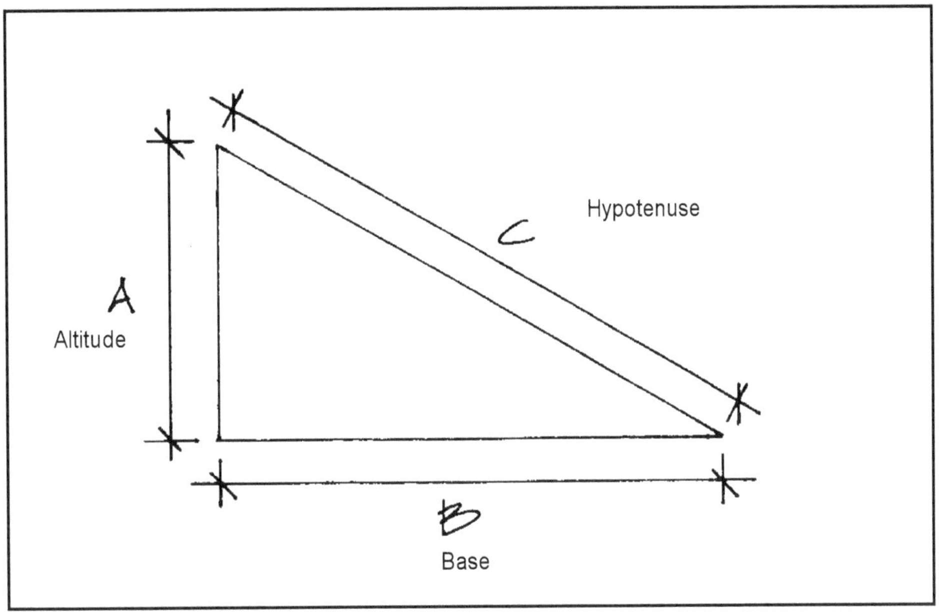

Figure 9.1 Right Triangle

Calculating the length of the hypotenuse is helpful in determining the length of gables at roofs and any diagonal dimension in a flat plane. To do so, the *Pythagorean Theorem* or the *Right Triangle Law* is used. The Right Triangle Law is the same as the "3-4-5 triangle" that framers and contractors use to "square up" work in the field. It states:

> *"The square of the hypotenuse of a right triangle is equal to the sum of the squares of the other two sides."*

Converting this to a formula:

> $C^2 = A^2 + B^2$, *where* C = *length of the hypotenuse,* A = *length of the altitude, and* B = *length of the base.*

This formula can be used to solve for the missing side of a right triangle, provided the lengths of the other two sides are known.

There are three types of triangles that a cost segregation practitioner will encounter in a set of construction drawings: an *equilateral* triangle has three sides that are equal in length. An *isosceles* triangle has two sides that are of equal length. In a *scalene* triangle, no sides are equal in length. *(See Figure 9.2.)*

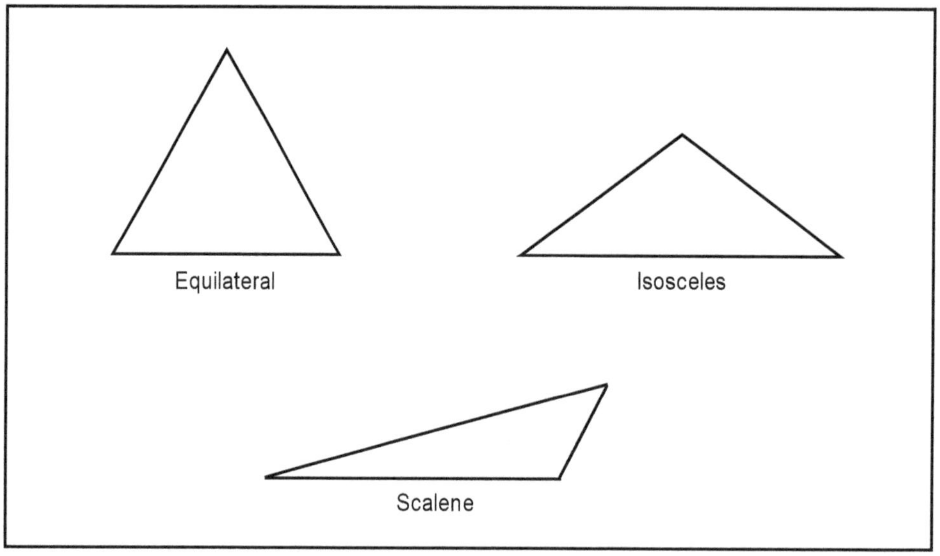

Figure 9.2

Area & Square Measure

Estimators are frequently required to determine the area of shapes. The most common shape in construction is some variation of the *rectangle*—four sides with all angles equal to 90°. Area calculations take into account only the surface, not the depth. Areas are expressed in square units (most commonly square feet, square yards, or square inches), the product of the length x width.

Triangular areas are calculated as:

> *A_T = ½ B x A, where A_T = area of a triangle, B = length of the base, and A = length of the altitude.*

Irregular Shapes

To calculate the areas of more complex polygons and other irregular shapes, you need to calculate several smaller areas and add and/or subtract. Rather than confuse the calculations with complicated formulas, it's simpler to calculate each recognizable shape, then add them all together. When determining the area of a portion of a larger (whole) area, deduct the unwanted areas until the desired area remains. For most construction applications, accurately approximating the area of an irregular shape is sufficient.

Area of a Circle

Many construction elements are circular, such as brick patios or concrete-filled sonotubes. Before reviewing the formula for the area of a circle, it is helpful to define the parts of the circle and some constant relationships.

The *circumference* is the perimeter of the circle. The *diameter* is defined as a line drawn through the center of the circle, beginning and ending on the circumference. Any number of diameters drawn on a circle should render two equal halves of that circle. All diameters of the same circle are equal.

The *radius* of a circle is a line from the center point within a circle to a point on the circumference. All radii of the same circle are equal in length. The radius, by definition, is one-half of the diameter. Knowing this relationship, one can establish the following formulas:

$D = 2 \times R$ or $R = 1/2 \times D$, where D = diameter and R = radius.

The circumference of a circle has a constant relationship to the diameter of 3.1416, "pi," with the corresponding symbol, π. For most construction calculations, π can be truncated at two places after the decimal point, or 3.14. If the area to be calculated is very large, π can be extended to a third decimal place, or 3.142. The formula for the relationship between the circumference and the diameter is as follows:

$C = \pi \times D$ or $C = 2 \times \pi \times R$, where C = circumference, $\pi = 3.14$, D = diameter, and R = radius.

A *chord* is a straight line connecting two points on the circumference, without passing through the center of the circle. An *arc* is any portion of the circumference of the circle.

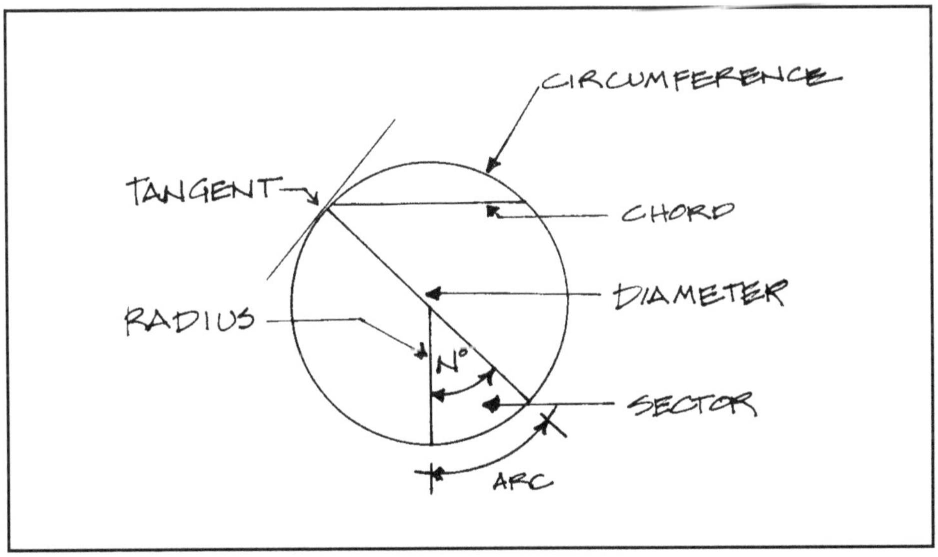

Figure 9.3

A circle has 360°. Therefore, if the radius and the interior angle between the radii are known, the length of the arc between them can be calculated using the formula:

$L_A = N/360 \times 2 \times \pi \times R$, *where N is the central angle,* $\pi = 3.14$, *and R = radius.*

The *tangent* of a circle is a straight line touching only one point on the circumference. A radius drawn to this point is at 90° to the tangent.

Using the mathematical relationships expressed above, it is possible to calculate the area of a circle if certain information about the circle is known.

The area of a circle is the radius multiplied by the radius then multiplied by π. As a formula:

$A_c = \pi \times R \times R$, *or* $A_c = \pi \times R^2$

An alternative method for calculating the area of a circle is to multiply the diameter by itself, then multiply the resultant area by the constant 0.7854.

$A_c = D \times D \times 0.7854$, *or* $A_c = D^2 \times 0.7854$, *where D = diameter and 0.7854 is a constant.*

The area of a portion of a circle (in a pie-piece shape, called a *sector*) can be calculated with a known radius and angle.

$A_s = N/360 \times \pi \times R^2$, *where* A_s *= area of a sector, N = angle between the radii in degrees,* $\pi = 3.14$, *and R = radius*

(For information on calculating the surface area of cylinders, pyramids, and cones, see Appendix C.)

Volume & Cubic Measure

In contrast to area, which has only two dimensions, volume has a third dimension—depth (also called *thickness* or *height*). The term *cubic* refers to the volume of a solid, whereas square accounts only for its area. The standard units of cubic measure are cubic inches, cubic feet (CF), and cubic yards (CY).

Volume of a Prism

Prisms (rectangles with three dimensions) in construction include items like pile caps, footings, and excavations. If the prism's dimensions of length, width, and height are the same, it would be a *cube,* with its volume measured in cubic feet (CF). Cubic feet are converted to cubic yards (CY) in concrete, where 1 CY = 27 CF.

The formula for the volume of a prism (a shape whose ends and opposite sides are parallel) is:

$V = A \times h$, *where V = volume of the prism, A = area of the base, and h = height.*

Or:

$V = l \times w \times h$ *where, l = length, w = width, and h = height.*

The triangular prism has a triangular surface area and a height. The rule of base area multiplied by height still applies. The volume of a triangular prism is:

$V = 1/2 \times l \times w \times h$, *where, again, l = length, w = width, and h = height.*

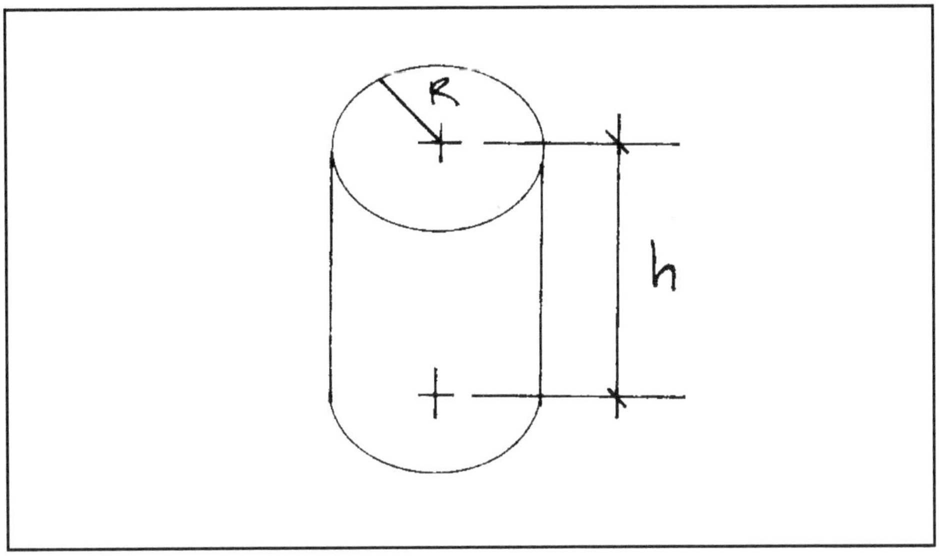

Figure 9.4

Volume of a Cylinder

Cylinders are a common shape in construction. For example, the formula for the volume of a cylinder is essential in calculating the amount of concrete to fill a sonotube or a round column form. The volume is the area of its circular base multiplied by its height:

> $V_c = \pi \times R^2 \times h$, *where* V_c *= volume of a cylinder,* $\pi = 3.14$, R *= radius, and* h *= height.*

Volume of a Pyramid

The volume of a pyramid, which commonly appears in the base of a tapered retaining wall, for example, is calculated as:

> $V_p = 1/3 \times A_b \times a$, *where* A_b *= area of the base of the pyramid provided that the base is a regular polygon, and* a *= the altitude of the pyramid as defined by a line drawn at 90° to the base to the vertex of the pyramid.*

Volume of a Cone

The volume of a cone is very similar to that of a pyramid:

> $V_c = 1/3 \times A_c \times a$, *where* A_c *= area of the round base of the cone, and* a *= the altitude of the cone as defined by a line drawn at 90° to the base to the vertex of the cone.*

Conclusion

The shapes in this section are by no means all the ones that could occur in a quantity takeoff as part of a cost segregation study, but they are the most common. It is important that an accurate takeoff be performed on the asset to be depreciated, and understanding the formulas for determining the measurement, area, and volume of those shapes is a starting point.

Chapter 10
Understanding Material & Labor Cost Factors

The engineering, or construction cost, portion of a cost segregation study involves compiling unit costs for building assets. To correctly apply unit costs, it is helpful to understand the components that comprise them. It is also important to understand the "adjustments" that construction professionals make in order to ensure accuracy in building cost estimates. Accurate costs include material, labor, and equipment, which are adjusted for factors such as location, taxes, material waste allowances, overhead (which includes such items as insurance, taxes, benefits, office expenses, and depreciation of equipment), and profit. This chapter will explain how construction cost estimators calculate and factor in all of these costs in order to convert a "raw" cost into a "burdened," or complete cost.

Materials

Two modifiers are routinely added to material costs to arrive at an accurate unit price: sales tax and waste. Most commercial construction projects are subject to sales tax on materials, and some states apply an additional tax to labor. Waste is applicable to most materials and is added to the installed quantity as a percentage increase to the unit price. The following example illustrates how carpeting for an office space would be priced, including waste.

The net area to be covered with carpet is 2,000 SY. The selected material is sold only in 15'-wide rolls. When layout is considered, the actual quantity to be purchased is 2,160 SY, or an additional 8% (allowing for waste), plus sales tax (5% in this example).

Carpet (as specified)	$38.00 per SY
Waste factor, 8%	$3.04 per SY
Subtotal	$41.04 per SY
Sales tax, 5%	$2.05 per SY
Total	$43.09 per SY

If the $43.09 is extended over the 2,000 SY of the net area, the cost will be $86,180. Profit also needs to be added to the unit price. If a 10% profit were added, the billing unit price would be $47.40 per SY.

(Note that since the correct quantity to actually be installed is 2,000 SY, it would be incorrect to use 2,160 SY (including waste) as the installed quantity for the labor portion of the estimate, since this would result in an 8% error.)

Labor

Calculating the cost of labor unit prices is more difficult than for material unit prices. First, the correct wage rate must be calculated. Modifiers such as state and federal taxes on wages, insurances, and benefits must be added to the cost of the actual wage rate.

While a cost segregation practitioner may not be required to calculate individual benefits, taxes, and insurance in order to produce an hourly labor cost for a construction estimate, practitioners may in some cases (such as an audit) require detailed information in order to substantiate a cost.

Billing Rate

The term *wage rate* for this discussion refers to the rate agreed upon between the employer and employee. Once the wage rate has been modified by taxes, insurance, and overhead, it is referred to as the *burdened labor cost.* After profit has been added, the result is the full value of the labor-hour, referred to as the *billing rate.* This is the amount that would be charged for labor in a time and material application. The following components make up the billing rate.

Wage Rate

The wage rate, or hourly rate of pay on which the employee's paycheck is calculated, is regulated by the agreement between the employee and the employer or a collective bargaining agreement. For example, a carpenter hired at $25.00 per hour would gross $1,000 in wages for a 40-hour work week. From this, taxes would be deducted, and the employee would receive a net paycheck.

Benefits

Many employers provide benefits in the form of medical and dental policies, vacation pay, paid sick days, retirement package contributions, annuities, or other forms of compensation. The costs to the employer of these benefits must be recovered, and are typically factored into the burdened rate. They can be broken down and expressed as a percentage of the hourly rate, then extended to a dollar amount per hour, which is added to the wage rate.

State and Federal Taxes/Insurance

An understanding of taxes and insurance as applied to wage rates may not be required for most cost segregation studies. However, there are cases when this information will come into play, such as a major developer or contractor performing a cost segregation study internally.

FUTA and SUTA Each state and the federal government apply a tax to wages to provide a source for unemployment, retirement, and disability benefits. The Federal Unemployment Tax Act (FUTA) and State Unemployment Tax Act (SUTA) are paid by employers based on employees' gross taxable wages. As of the printing of this book, the FUTA is at 0.8% (6.2% less the maximum state credit of 5.4% of SUTA paid) with a $7,000 cap. Once the employee earns more than the cap, the tax stops.

The SUTA tax varies from state to state and will fluctuate depending on the employer's experience, which is measured by the number of employees who have been laid off. The fewer employees claiming benefits, the less the employer's contribution. As the number of employees claiming benefits increases, so does the employer's contribution. The SUTA tax varies based on a number of factors, including the employer's experience in the work force, the number of workers, and type of work. There is typically a cap on wages, as with the FUTA tax.

FICA The Federal Insurance Contribution Act (FICA) (also referred to as Social Security and Medicare) is defined by the federal government, but more appropriately considered insurance than a tax. Employees are taxed at the rate of 7.65% of gross taxable wages. Additionally, the employer makes a matching contribution of the same percentage. It is this percentage that is recouped in the wage calculation.

Worker's Compensation Insurance Another type of insurance to protect the employee, compulsory in most states, is Worker's Compensation. It is provided by employers in the event of a worker's injury, disablement, or death occurring in the workplace. Workers' Comp is different for each worker or trade classification, and is based on a percentage of the employee's gross non-premium wages. (Overtime or premium wages do not affect the insurance premium.) Insurance premiums for individual companies performing the same work vary by experience—i.e., premium rates are based on the employer's number of days without a serious accident on the job site. A safe work environment translates to lower Workers' Compensation premiums. From an estimating point of view, lower premiums are reflected in lower labor costs, thereby enabling contractors to bid more competitively.

General Liability Insurance General liability insurance protects the project and any adjacent properties from damage. This insurance compensates the contractor and, ultimately, the owner for damage by parties under the general contractor's control, including employees and subcontractors. Like Worker's Compensation, general liability premiums are based on the contractor's success in reducing risk and are calculated based on two factors: employee wages and subcontracted work. For most general contractors who subcontract more than 70% of their contract work, the largest portion of the premium is based on the costs incurred under subcontract. However, a portion of the premiums is based on gross employee wages. Percentages vary with the policy limits, size of the job, contractor's performance history, and the state's or insurer's regulations.

Indirect Overhead

The IRS *Cost Segregation Audit Techniques Guide* provides that a portion of the contractor's indirect overhead costs can be allocated to a particular asset. Chapter 4 of the guide states that the preparer should explain, within the study, how the indirect costs were allocated to the particular asset. Indirect costs, as applied to a construction cost estimate, are as follows.

Costs related to the operation of the contractor's main office and staff are called *indirect overhead* expenses. These costs are associated with maintaining a business, but are not directly attributable to one specific project. Indirect overhead includes, but is not limited to, expenses such as:

- Corporate officers' salaries and benefits
- Rent or mortgage on the main office
- Monthly telephone, fax, or Internet charges
- Clerical and office associates' salaries, insurances, and taxes
- Corporate vehicles, and their insurance and maintenance/operating costs
- Estimator's salary and benefits
- Legal, accounting, and technology service fees outside the firm
- Heat, electricity, and maintenance of the main office

To recover these costs, a small amount for each is added to each project, as a percentage of the cost of the work.

Profit

Profit is the reason for doing business—the end result of a project done right, and/or the reward for risks taken. Profit needs to be captured in the calculation of the billing rate. Profit is most often assigned as a percentage of all costs for the work, but it can also be assigned as a lump sum or stipulated fee. Calculating profit is based on careful consideration of a host of factors, some less tangible than others. For the examples in this text, we will assume profit is 10% of costs.

Below is a sample calculation to illustrate the billing rate for a carpenter, based on the wage rate and using hypothetical rates for insurance, taxes, and other markups.

Carpenter wage rate	$25.00 per hour
Benefit package	8.55
FICA (7.65% on wage)	1.92
FUTA (0.8% on wage)	0.20
SUTA (7% on wage)	1.75
Worker's Compensation at carpenter rate of 11.05% (on wage)	2.77
General liability insurance at the rate of 0.82% on wage	0.21
Subtotal	$40.40
Indirect overhead at 10% of all costs	4.04
Subtotal burdened labor cost	$44.44
Profit at 10% of all costs	$4.44
Billing rate	$48.88 per hour

Labor Costs and Productivity

Productivity can best be described as the rate at which work is performed by an individual or crew per unit of time—usually per day. A day is considered an eight-hour period, over which the productivity will fluctuate up or down, depending on breaks, learning curves, starts and stops, and material handling. When the work produced is measured at the end of the day, it represents a reasonable average for all of the normal fluctuations. It is fairly easy to take a "snapshot" at the beginning of the day and again at the end of the day and compare the difference.

Productivity on the same task can vary when performed by two different crews, during different times of the day, or with extremes in weather or temperature. There are many other factors to consider when determining the productivity of a task. For example, consider two separate roofs with the same number of units (squares), height from the ground, asphalt shingles, and configuration (style). The only difference is the pitch; one is 6:12, and the other is 12:12. Most construction professionals would agree that the steeper-pitched roof reduces mobility, thereby making the roofing process slower and more difficult. This is translated into a reduced productivity, and a higher unit cost for labor.

The best measure of a crew's productivity is a company's own historical data. Job cost records for payroll and labor-hours to complete a specific task offer the best guideline for establishing performance, or how much of a task can be completed per day. Experienced professionals know that while this is no guarantee, historical data provides a logical model. Many of the published cost data sources used in pricing assets are based on a designated "average" productivity, which may need to be adjusted to match the circumstances of the project being evaluated. Projects with difficult aspects or logistics, such as a downtown location, may be more expensive, and may need to be adjusted to reflect the difficulties. Projects with below-average costs can also occur in some situations.

Crew and Individual Tasks

All tasks can be categorized for estimating purposes as individual or crew. Individual tasks are performed by a single worker, with performance and productivity measured by the output of one individual. For example, a carpenter's performance, trimming interior windows with wood casing, can be measured by counting the number of windows trimmed in a single day, provided the windows are of similar size. After several days' work, an average productivity rate could be established, and a model created for future window trimming work. The cost of the carpenter per day could then be divided by the number of windows trimmed to establish a baseline cost model per window. For example:

> *8 carpenter hours x $48.88 per hour = $391.04. For multiple days, the carpenter will trim 10.5 windows per day. The cost per window is $37.24. It could also be concluded that approximately 0.76 carpenter hours are needed to trim a window.*

(Adjustments could be made to this cost model to establish the productivity for trimming windows of different sizes.)

Estimating crew task productivities is more challenging because crew tasks combine the performance of different component functions by different individuals. Each performs a specific function to complete the overall task. The productivity rate is a result of how well the team performs together.

For example, a crew installing brick work might consist of two mason tenders and three bricklayers, including a foreman. Each member has a specific function that contributes to the overall performance of the crew. The foreman directs the crew, lays out the work, and aspires to a steady, predictable productivity. The bricklayers, including the foreman, lay up the brick. The mason tenders (laborers tending the bricklayers) mix mortar, stock the staging with materials, and provide support services. While they do not install the brick, their duties are essential to successful production. Each crew member's performance relies on other members to perform their work efficiently. Productivity rates are based on the crew, not just the individual. For a larger crew on more complex tasks, a foreman or on-site supervision must also be calculated as part of the cost of the work.

Subcontractors

Subcontractors are, by definition, independent contractors. They "hire and fire" their own workers, purchase their own materials, run their own equipment, and define their success based on their own performance. Subcontractors provide their own insurances—both general liability and Worker's Compensation. They are held responsible for their own actions and are required to pay their own taxes. By virtually every test, they are independent of the general or prime contractor.

The major portion of the average commercial construction project is performed by subcontractors. This arrangement is supported by the breakdown of line items in the AIA G702 and G703, Application for Payment forms, for projects that have been completed or are in the process of being constructed. For the work of subcontractors, all of the costs for burden and markups on materials and labor, as well as direct and indirect overhead and profit, have been included in the price. This is an important point for determining the final unit price to be used in a cost segregation study. However, the general contractor's indirect costs and profit need to be added on top of the unit prices from subcontractors, since this is the cost recognized in the Schedule of Values.

A Final Word on Labor Costs in Cost Segregation

It is important to note that one of the major differences between cost segregation and conventional estimating is that the cost segregation practitioner in most cases begins work *after* all costs have been incurred and/or the building construction is complete. The conventional estimator estimates *before* construction starts. Exceptions include projects that are completed in phases over more than one tax year, or reports generated for tax planning on projects, which are conceptual, or in the planning stages. As a result, detailed labor cost calculations rarely appear in cost segregation reports, which include labor and material unit costs (and items they incorporate) referenced through a costing source such as RSMeans. These costs already include overhead and profit.

The cost segregation practitioner, in most cases, does not have access to the contractor's internal rates as applied to that project, and may not have the time to work with the contractor to establish this level of detail. The contractor often covers general conditions and overhead costs as lump sum values in new construction. Without available detail, these costs are treated as indirect and are allocated on a pro-rata basis across all assets.

When generating reports, it is important to note whether union or open shop labor was used on the project, so that appropriate unit costs can be assigned. Also, it must be determined whether overhead costs need to be added to the unit cost, or if they are already part of indirect costs and being allocated. This is important to avoid including costs twice, or leaving them out altogether.

Conclusion

An understanding of the components and factors that go into unit prices strengthens the credibility of a cost segregation study. Many costs must be modified from the "base" costs before they can be applied accurately to an asset. Others, such as subcontractors' costs, have the modifiers included and require no changes. A cost segregation practitioner will, in most cases, not need to define or document unit prices to the level of detail explained in this chapter. However, complex cost segregation studies, and increasing quality expectations from owners—in terms of complete and defensible work—make it important to understand the basis behind accuracy in construction costs.

Chapter 11

Sources of Pricing Information

Cost segregation studies require accurate, properly documented information to substantiate the cost of depreciated assets. Well-prepared and organized cost segregation studies, complete with documented cost information, present less of a challenge to examiners, and this may, in turn, expedite the review process. This chapter will present various types and levels of detail of pricing information, as well as information sources and their reliability.

Estimates that Use Unit Prices

Each and every element within an estimate—from quantities to unit prices—must be substantiated in factual terms in order to be considered professional work. This is the basis for a defensible, detailed estimate. Done correctly, the *unit price estimate* assigns an accurate quantity and price to each item of work or asset identified in the study. Unit prices reflect the material and labor costs of specific components based on commonly used units of measure. When the cost of the individual components of an asset are summarized, and indirect costs are added, the result is referred to as the *allocated cost.*

The following basic parameters define a unit price:

- Unit prices are based on dollars or portions of a dollar per unit.
- The pricing unit, or unit cost, should be incorporated into the extended cost based on the task quantities.
- Many unit prices are based on a specific quality of construction.
- Labor unit prices will vary in cost based on "open shop" or union pay scales.
- Unit prices can be based on actual, documented project cost data or published cost data.
- Unit prices can include all, one, or a combination of cost types (material, labor, equipment, subcontractor, etc.).

Cost Data Sources

Collecting, organizing, and analyzing the cost data for a cost segregation study is no small task. The ability to make efficient use of this data depends on how well it has been organized. Cost data can come from several sources, including:

1. Requisitions or applications for payment submitted by the general or prime contractor.
2. Requisitions or applications for payment submitted by subcontractors.
3. Invoicing from consultants, vendors, and suppliers outside of the general contractor's payment request.
4. Cost data from published sources.
5. Actual estimates from manufacturers, suppliers, or contractors that were not involved in the project. (While these are acceptable sources, they may be reluctant to provide this information, as there is little or no gain for them.)

Categories 1-3 are referred to as *contemporaneous* pricing sources and are based on a review of the actual project's invoicing documents, plans, and specifications. These costs can be directly linked to the outlay of capital for the construction project. Category 4 costs, from published cost data, such as RSMeans, can be particularly useful when actual prices are not available for the project.

Contemporaneous Pricing Sources

The most common source of contemporaneous pricing for new construction is the general contractor's requisition form. The most widely accepted and used requisitioning format is the American Institute of Architect's documents G702 and G703, together they are formally called the "Application for Payment." These forms provide for an itemized breakdown of direct and indirect construction costs for a particular project and reflect the total contract amount.

These AIA forms are designed to "apply for," or *requisition* for, partial progress payments based on work completed at the end of the pay period (usually one month). Each monthly requisition is acknowledged in the billing that follows, by a cumulative accounting process. This method allows the cost segregation analyst to establish which portion of a particular asset or task was completed in a specific month or larger time period. For recently completed construction, this document should be readily available (for each billing period) within the owner's collection of records. Change orders may also be included as part of the normal payment process.

The contract total is broken down into individual work tasks or classifications of work that can be any combination of material, labor, equipment, subcontract, or other expense. Work tasks are itemized on a line-by-line basis in a "Schedule of Values," which lists the dollar value of each task at 100% complete. Assets can be composed of individual line items, multiple line items combined, or a portion of an individual line item. Figure 11.1 is a sample generic Application for Payment form, used in the same way as the AIA G702 and G703 forms.

PAYMENT APPLICATION — Page 1 of 2

PROJECT: XYZ Manufacturing Co
New Manufacturing Facility for XYZ

APPLICATION #:
DATE OF APPLICATION: (DATE)
PERIOD THRU: (DATE)
PROJECT #s: XYZ-1

Payment Application with Contractor's signature attached.

A	B	C	D	E	F	G		H	I
			COMPLETED WORK						
ITEM #	WORK DESCRIPTION	SCHEDULED AMOUNT	AMOUNT PREVIOUS PERIODS	AMOUNT THIS PERIOD	STORED MATERIALS (NOT IN D OR E)	TOTAL COMPLETED AND STORED (D + E + F)	PERCENT COMPLETE (G / C)	BALANCE TO COMPLETION (C-G)	RETAINAGE (IF VARIABLE)
1	Building Permits and Fee	$4,000.00							
2	Performance and Payment Bond	$12,000.00							
3	Prepaid Insurance Premiums	$2,500.00							
4	Project Overhead and General Con	$35,000.00							
5	Mobilization	$10,000.00							
6	Site Clearing and Preparation	$12,000.00							
7	Excavation and Backfill Foundation	$23,000.00							
8	Paving and Curbing	$16,000.00							
9	Drainage & Structures	$23,000.00							
10	Water & Sewer Utilities	$26,000.00							
11	Finish Grading of Site	$8,000.00							
12	Landscape and Irrigation	$14,000.00							
13	Site Improvements	$17,500.00							
14	Foundation and Slab	$37,000.00							
15	Structural Steel, Joist and Deck	$88,900.00							
16	Exterior Curtain Wall Assembly	$93,700.00							
17	Roof System and Insulation	$54,000.00							
18	Skylights	$8,000.00							
19	Doors and Hardware	$12,300.00							
20	Windows	$47,000.00							
21	Interior Casework and Trim	$36,400.00							
22	Wainscoting & Trim at Conference	$17,800.00							
23	Insulation at Walls & Ceilings	$12,300.00							
24	Drywall & Tape & Finish	$69,400.00							
25	Interior Painting	$12,700.00							
26	Acoustical Ceiling Systems	$14,600.00							
27	Epoxy Flooring	$33,000.00							
28	Direct Glue Carpet	$12,000.00							
29	Heavy Duty Metal Lockers	$8,000.00							
	SUB-TOTAL	$760,100.00							

Figure 11.1

PAYMENT APPLICATION — Page 2 of 2

PROJECT: XYZ Manufacturing Co
New Manufacturing Facility for XYZ

APPLICATION #:
DATE OF APPLICATION: (DATE)
PERIOD THRU: (DATE)
PROJECT #s: XYZ-1

Payment Application with Contractor's signature attached.

A	B	C	D	E	F	G		H	I
			COMPLETED WORK						
ITEM #	WORK DESCRIPTION	SCHEDULED AMOUNT	AMOUNT PREVIOUS PERIODS	AMOUNT THIS PERIOD	STORED MATERIALS (NOT IN D OR E)	TOTAL COMPLETED AND STORED (D + E + F)	PERCENT COMPLETE (G / C)	BALANCE TO COMPLETION (C-G)	RETAINAGE (IF VARIABLE)
30	Manufacturing Benches	$8,400.00							
31	Shelving	$15,000.00							
32	Fire Protection Systems	$67,900.00							
33	Plumbing -Domestic Services	$24,600.00							
34	Plumbing -Manufacturing Process	$78,600.00							
35	HVAC Systems-Office Area	$54,000.00							
36	Ventilation System-Manufacturing	$67,500.00							
37	Heating Systems-Manufacturing	$78,000.00							
38	Electrical Power-Office	$8,000.00							
39	Electrical Power-Manufacturing	$16,700.00							
40	Lighting-Office & General Manufact	$27,000.00							
41	Lighting-Specialty Task at Manufact	$44,000.00							
42	Site Lighting	$22,800.00							
43	Signage & Sign Lighting	$13,000.00							
	TOTAL	$1,285,600.00							

Figure 11.1 continued

Billings Other than the General Contractor

Not all project costs include a Schedule of Values. Some costs originate outside of the general contract and are invoiced directly to the owner. (Most contracts allow the owner to contract separate scopes of work outside of the prime contract with the general contractor.) Under these conditions, the owner will incur invoices from other parties, such as from independent consultants, vendors, suppliers, or other contractors. Cost segregation practitioners should review all project invoicing for any category of work performed, equipment purchased, or services rendered for a particular asset.

Published Cost Data

Contemporaneous pricing sources are often unavailable for use in cost segregation studies, particularly for buildings that were constructed or acquired in the past. Lost or incomplete records, uncooperative participants in the construction process, or any number of situations may prevent cost segregation practitioners from using actual costs to reconstruct the asset cost. In such cases, alternative pricing from a recognized, published source such as RSMeans, must be used to produce a professional cost segregation study.

RSMeans cost data, which is substantiated by a comprehensive reference section, includes factors that can be used to adjust costs to specific locations. RSMeans is specifically recommended by the IRS in its *Cost Segregation Audit Techniques Guide* as a standard for accurate cost data. As previously noted, comprehensive cost segregation studies generally rely on unit price data, which is available from RSMeans in both printed form and computerized databases.

As with any published construction cost data source, it is essential that users understand the parameters and conditions that have been applied in compiling it. Published cost data is typically based on averages and will vary with factors such as location, climate, skill of the crew, availability of resources, and supervision, and it must be adjusted for these considerations. Published data is also based on normal working conditions, during regular business hours, and under average conditions. Unpredictable costs associated with labor resource shortages, supply and demand cycles, unusually harsh weather conditions, or travel and per diem costs have to be calculated separately.

Published cost data can be presented in two ways: as *bare costs* without overhead and profit, or as *total costs* including a markup for overhead and profit. It is important for practitioners to understand whether the work was self-performed (by the general contractor's own employees) or subcontracted. If self-performed, bare costs would be preferable so that the markup can be applied to the estimate total. If the work is subcontracted, an allowance for overhead and profit for the installing contractor must be carried in the estimate before recapitulation.

RSMeans cost data lends itself particulary well to use in a cost segregation study. Organized according to the 16 divisions of the CSI MasterFormat, it provides a task description and a line item number for referencing in a cost segregation study, a daily output by a specifically defined crew, and the units of each task. Each unit price is broken down into its material, labor, and/or equipment components without markups for overhead and profit; a separate column adds these markups. The data can be used to recreate the costs of an asset at any particular date—current or in the past. Cost segregation analysts can review the plans, specifications, and addenda; perform the takeoff on the qualifying assets; and apply the unit price information from the appropriate data source. Figure 11.2 shows an example of RSMeans 2005 cost data.

To demonstrate the process of applying published data, consider the following example.

> *Assume that a cost segregation analyst has detemined that a property has directly adhered (glued down) carpet that qualifies for accelerated depreciation. The analyst has taken off the quantity from the plans and detemined that the net quantity for the purpose of study is 3,000 SY. He or she has also consulted the appropriate specification section (unit price section 09680) in Division 9 – Finishes, and determined that the carpet is a commercial-grade, direct cement, 26 oz. nylon, level loop. A search of RSMeans 2005* Building Construction Cost Data, *section 09680–Carpet, reveals that line item 09680-800-0700 is a sufficiently close match to the specified product. The cost per SY, including overhead and profit, is $23. If the total quantity of 3,000 SY is multiplied by $23 per SY, the extended price and value of the asset is $69,000. (See Figure 11.2.)*

This same procedure can be done electronically using Means *CostWorks,* which can export data to a spreadsheet.

Location Factors

Costs vary depending on regional economy. The national average costs in RSMeans publications can be adjusted easily to precise locations by using the more than 950 location factors for major cities throughout the U.S. and Canada.

How to Use Line Items

Each "line item" is assigned a specific 12-digit line item number, which identifies its CSI division, subdivision, and major classification. The descriptions are arranged in a hierarchical format. If an item is indented farther to the right than any item above it, some portion of the above description applies. Some part of each of those lines makes up the complete description for the item. Referring to the main descriptions and the sub-descriptions that follow, an appropriate, complete description can be created. If additional information is needed, or if it is necessary to modify the published information in some way, all the components needed to add to or modify the cost are available in the columns to the right of the description.

To develop a unit cost, certain pieces of information are necessary, including who has performed the work; what equipment was needed; how much work could be completed in a set time period; and how much the crew, equipment, and materials cost. Simply put, a unit cost is developed like this:

> *A carpenter earns $200 per 8-hour day. That carpenter can install 200 LF of wood trim per day. The wood trim costs $1.00 per LF. The total unit cost for installed wood trim would be:*
>
> *$200 per day/200 LF per day = $1.00 per LF for labor, plus: $1.00 per LF for the material, or $2.00 per LF to supply and install the wood trim.*

This formula is the basis for all unit price line items published by RSMeans. For example, in the carpet example referenced earlier in this chapter, all the information required to develop the cost is in the columns of information provided. The "Crew" column indicates type(s) of worker(s) needed to perform the work and what equipment (if any) is required.

09600 | Flooring

		09673 | Composition Flooring	CREW	DAILY OUTPUT	LABOR-HOURS	UNIT	2005 BARE COSTS MAT.	LABOR	EQUIP.	TOTAL	TOTAL INCL O&P	
100	0200	Methyl methachrylate, 1/4" thick	C-8A	3,000	.016	S.F.	5	.47		5.47	6.20	100
	0210	1/8" thick	"	3,000	.016		2.75	.47		3.22	3.74	
	0300	Cupric oxychloride, on bond coat, minimum	C-6	480	.100		2.77	2.81	.10	5.68	7.50	
	0400	Maximum		420	.114		4.63	3.21	.11	7.95	10.15	
	0600	Epoxy, with colored quartz chips, broadcast, minimum		675	.071		2.19	2	.07	4.26	5.55	
	0700	Maximum		490	.098		2.66	2.75	.10	5.51	7.25	
	0900	Trowelled, minimum		560	.086		2.83	2.41	.09	5.33	6.90	
	1000	Maximum		480	.100		4.12	2.81	.10	7.03	8.95	
	1200	Heavy duty epoxy topping, 1/4" thick,										
	1300	500 to 1,000 S.F.	C-6	420	.114	S.F.	5.05	3.21	.11	8.37	10.60	
	1500	1,000 to 2,000 S.F.		450	.107		3.99	2.99	.11	7.09	9.10	
	1600	Over 10,000 S.F.		480	.100		3.69	2.81	.10	6.60	8.50	
	1800	Epoxy terrazzo, 1/4" thick, chemical resistant, minimum	J-3	200	.080		5.30	2.35	1.05	8.70	10.45	
	1900	Maximum		150	.107		8.20	3.13	1.40	12.73	15.15	
	2100	Conductive, minimum		210	.076		6.60	2.24	1	9.84	11.65	
	2200	Maximum		160	.100		8.95	2.94	1.31	13.20	15.60	
	2400	Mastic, hot laid, 2 coat, 1-1/2" thick, standard, minimum	C-6	690	.070		3.21	1.95	.07	5.23	6.60	
	2500	Maximum		520	.092		4.12	2.59	.09	6.80	8.60	
	2700	Acidproof, minimum		605	.079		4.12	2.23	.08	6.43	8.05	
	2800	Maximum		350	.137		5.70	3.85	.14	9.69	12.35	
	3000	Neoprene, trowelled on, 1/4" thick, minimum		545	.088		3.16	2.47	.09	5.72	7.40	
	3100	Maximum		430	.112		4.35	3.13	.11	7.59	9.75	
	3150	Polyacrylate terrazzo, 1/4" thick, minimum		735	.065		2.98	1.83	.07	4.88	6.15	
	3170	Maximum		480	.100		3.85	2.81	.10	6.76	8.65	
	3200	3/8" thick, minimum		620	.077		3.92	2.17	.08	6.17	7.75	
	3220	Maximum		480	.100		5.75	2.81	.10	8.66	10.80	
	3300	Conductive terrazzo, 1/4" thick, minimum		450	.107		6.95	2.99	.11	10.05	12.35	
	3330	Maximum		305	.157		8.90	4.42	.16	13.48	16.75	
	3350	3/8" thick, minimum		365	.132		9.30	3.69	.13	13.12	16	
	3370	Maximum		255	.188		11.45	5.30	.19	16.94	21	
	3450	Granite, conductive, 1/4" thick, minimum		695	.069		7.25	1.94	.07	9.26	11.05	
	3470	Maximum		420	.114		9.35	3.21	.11	12.67	15.35	
	3500	3/8" thick, minimum		695	.069		10.60	1.94	.07	12.61	14.70	
	3520	Maximum		380	.126		12.85	3.54	.13	16.52	19.70	
	3600	Polyester, with colored quartz chips, 1/16" thick, minimum		1,065	.045		2.89	1.26	.05	4.20	5.15	
	3700	Maximum		560	.086		4	2.41	.09	6.50	8.20	
	3900	1/8" thick, minimum		810	.059		3.17	1.66	.06	4.89	6.10	
	4000	Maximum		675	.071		4.47	2	.07	6.54	8.05	
	4200	Polyester, heavy duty, compared to epoxy, add		2,590	.019		1.21	.52	.02	1.75	2.15	
	4300	Polyurethane, with suspended vinyl chips, minimum		1,065	.045		6.15	1.26	.05	7.46	8.80	
	4500	Maximum		860	.056		8.85	1.57	.06	10.48	12.20	

		09680 | Carpet	CREW	DAILY OUTPUT	LABOR-HOURS	UNIT	MAT.	LABOR	EQUIP.	TOTAL	TOTAL INCL O&P	
600	0010	**CARPET PAD**, commercial grade										600
	9000	Sponge rubber pad, minimum	1 Tilf	150	.053	S.Y.	3.18	1.74		4.92	6.05	
	9100	Maximum		150	.053		8.25	1.74		9.99	11.65	
	9200	Felt pad, minimum		150	.053		3.57	1.74		5.31	6.50	
	9300	Maximum		150	.053		6.70	1.74		8.44	9.95	
	9400	Bonded urethane pad, minimum		150	.053		3.95	1.74		5.69	6.90	
	9500	Maximum		150	.053		6.75	1.74		8.49	10	
	9600	Prime urethane pad, minimum		150	.053		2.27	1.74		4.01	5.05	
	9700	Maximum		150	.053		4.19	1.74		5.93	7.15	
800	0010	**CARPET** Commercial grades, direct cement										800
	0700	Nylon, level loop, 26 oz., light to medium traffic **CN**	1 Tilf	75	.107	S.Y.	16.45	3.49		19.94	23	

Figure 11.2

The "Daily Output" column shows how much work the crew can complete in an 8-hour day. The "Labor-Hours" state how long it takes to install one unit of work (e.g., one square yard of carpet). The "Unit" column provides the unit of measure for everything found in that particular line item. Finally, the "Material," "Labor," and "Equipment" columns list each of those costs.

Material and labor costs are based on RSMeans industry research. Labor is derived by dividing the cost per day for the crew by the daily output. Each of the trades is categorized by individual tradespeople, such as carpenters (abbreviated "Carp"), and defined crews (such as "F-6," made up of multiple carpenters, laborers, and equipment operators). The daily costs for each of the trades can be found in the reference section of the cost data. Equipment costs, derived by dividing the cost for the equipment by the daily output, are also found in the reference section.

The "Total" column is the mathematical sum of the material, labor, and equipment columns. The "Total Incl. O&P" column is the "Total" unit price with overhead and profit added in. An amount equal to 10% is added to the material and equipment costs. Depending on the trades involved, an amount of up to 101.3% is added to the labor costs for Worker's Compensation insurance, taxes and insurance, home office overhead, and profit. This information will allow cost segregation practitioners to make any alterations or modifications that are required.

Conclusion

Whenever available, actual costs are always preferable as a basis for costing an asset. Unfortunately, this information is not availabe in the case of many cost segregation studies. Professional, published cost data is, however, a reliable and accepted resource for deriving asset costs in cases where actual costs are not available. In using published data, some construction knowledge and experience will allow cost segregation analysts to move through the data more quickly, find the items needed, and include them in the study.

As many as 50% of cost segregation studies may be completely without actual costs or contemporaneous cost data, thereby requiring that all of the costs be replicated through the use of published information. Historical factoring, such as RSMeans Historical Cost Indexes, can "deflate" costs, bringing them back to the time when the asset was placed in service. The cost segregation practioner also needs to modify or adjust cost estimates appropriately to account for numerous situations such as partial ownership, step-ups in basis, like-kind exchanges, bargain purchases, or other situations.

Chapter 12 Cost Segregation in Design & Construction

It is clear, based on the tasks described in the previous chapters, that a quality cost segregation study is a comprehensive, complex research document—one that usually requires the input of multiple parties with diverse professional backgrounds. Often, it is the owner's accounting professional who suggests the cost segregation study and goes on to orchestrate it, acting as the "point person" for the team that will need to be assembled to complete it.

Even the most elementary cost segregation studies may involve architects, engineers, consultants, general contractors, subcontractors, or major vendors and suppliers, who provide project documents and/or cost estimates. The owner's representative will provide coordination between the parties and access to both existing records and the facility itself (for site inspection). While these individuals are commonly called upon to contribute to cost segregation studies in the aforementioned ways, there are additional opportunities for each of these parties to actively offer and promote their own services to support cost segregation—starting in the design phase of new construction projects.

Cost Segregation in the Pre-Construction Phase

While most cost segregation studies involve properties that have already been acquired or built and are currently in use, this is not always the case. As cost segregation becomes more popular, an increasing number of studies are planned before a building or major renovation is designed, or construction begins.

There are significant advantages to implementing cost segregation early in the process. The design team can help maximize the owner's tax benefits if they are aware of asset depreciation guidelines as they select the building systems. Advance planning and organization for cost segregation can also alleviate or prevent some of the most common problems, such as incomplete documentation, that tend to arise if the process is not begun until after construction has been completed.

One of the biggest challenges is acquiring project cost information in a usable format. Although construction costs may be broken down on the AIA G702 and G703 forms, the Application for Payment, there are typically questions that still need to be answered.

Cost segregation practitioners who are performing a study on a building yet to be built should get involved at the earliest possible stage. In an ideal situation, they would contribute in the early phases of design development, when they can convey to architects, engineers, and consultants not only the principles and benefits of cost segregation, but guidelines for proper collection of data and specific building systems in terms of their tax advantage.

If owners are contracting with a general contractor by negotiation versus "hard bid," early planning for cost segregation gives them the opportunity to request that the contractor separate out all costs related to particular, specified assets. This can be done on a value-added comparison basis—for example, comparing the initial cost of a dedicated electrical circuit for specific display lighting slated for recovery through accelerated depreciation, versus the cost if the circuit is shared with other, more general types of lighting (and does not qualify for accelerated depreciation). The analyst can make suggestions for design features or building systems based on prior experience and/or IRS regulations or Tax Court case precedents. Various pricing scenarios can be reviewed, and a course of action can be decided, based on the best value of the investment over time.

The Cost Segregation Professional's Role

During the preliminary design phase, each of the team members has specific responsibilities and duties related to a cost segregation study. The cost segregation practitioner who is leading the study must:

- Become educated about the function of the structure to be built, understanding the business and how the structure is being built to suit the special requirements of that business.
- Review IRS regulations and Sections 1245 and 1250 property classifications to determine which particular assets will qualify for recovery in the specific application.
- Prepare a detailed list of the *anticipated assets* scheduled for recovery, and define the parameters of the asset for other team members.
- Review with architects, engineers, consultants, and the owner the specific anticipated assets, and how they will qualify for recovery. Particular focus should be on the *special conditions* that make each asset qualify as 1245 and 1250 property.
- Avoid design features that could lead to a challenge during an IRS audit. For example, electrical power distribution should have separate panels for standard equipment such as HVAC versus specialized manufacturing equipment that is used to support the particular trade or business.
- Select the cost segregation methodology up-front. While it may seem obvious that the detailed engineering approach based on actual cost records is preferable and the most likely option, accurate cost records do not happen by accident. *(See Chapter 6 for more on cost segregation methodologies.)* Detailed breakdowns for the Schedule of Values should be established, with specific line items identified for costing itemization. This task has particular applicability to architects, as they are most often the ones to review and approve the Schedule of Values and, ultimately, each of the contractor's applications for payment.

The cost segregation analyst should take these steps as the project gets under way.

- **As design development proceeds:** Periodically review the documents to ensure the anticipated assets have been detailed clearly in the plans and specs.
- **At the final stage of design:** Prior to issuing the bid set of documents, request that the architect and engineers provide a takeoff of the specific quantities that constitute each asset.
- **After the bid process and contract award:** Review the cost segregation process in detail with the selected general contractor.
- **During the construction process:** Review all supporting documentation accumulated during the construction process—invoices, applications for payments, change orders, and similar mechanisms that support/help define the cost of the asset.
- **At project completion:** Review the as-built documents in comparison with the original documents. Identify major changes from the original documents.

The Design Team: Architects & Engineers

Similar to accounting professionals, architects and engineers are always looking for the value-added feature or service that can distinguish their firm from the competition. Cost segregation provides this opportunity. Offering tax-oriented design and proper support documentation that helps the owner maximize an investment in new construction or remodeling is clearly a valuable service that can expand a design professional's repertoire. The architect's or engineer's role in cost segregation can have several facets.

For lack of a better term, "cost-seg friendly," or "tax-oriented," design can be applied to almost any commercial building venture without sacrificing the architectural aesthetics. The retail industry, where cost segregation has gained enormous acceptance, has been a leader in making cost-seg friendly design a reality. Many "big box" retailers have taken the added step of directly purchasing much of the major equipment, such as HVAC units, electrical light fixtures, store display fixtures, and the like. In addition to obtaining discounts for large purchases, this approach enables owners to secure the invoicing that is needed for contemporaneous documentation. With future plans for "roll-outs" of additional stores, past cost records can also be used to produce models for budgeting.

The design professional providing cost segregation support would take the following actions:

- Review and understand the assets designated for recovery under the guidelines set by the cost segregation practitioner.
- Whenever possible, design systems or assemblies for accelerated-depreciation assets so that they "stand alone," or are independent of other, more general systems. For example, provide circuitry and wiring for display lighting (that qualifies as personal property)—separate from that of general lighting (which does not qualify).

- Specify in the project manuals provided to the contractor all assets that must be priced and identified individually in the AIA Schedule of Values.
- Provide alternate designs (value engineering) for the assets that qualify for accelerated recovery and have the largest intial cost. (Value engineering is fairly standard practice for any building systems that have a large cost.)
- Specify products or equipment that qualify for recovery under IRS regulations. Many items, such as demountable partitions and cubicle compartments, are designed and built specifically to qualify as personal property. Retain any supporting documentation (such as submittals or product literature that identify products as depreciable through an accelerated tax recovery process) for inclusion within the study.
- Periodically review the design of identified assets with the cost segregation professional, and do so again prior to release of the documents.
- Implement a system for tracking and identifying any change orders that will have an impact on the cost segregation study.
- Review all supporting documentation provided by the contractor or subcontractors to ensure that it is complete and sufficiently detailed to qualify as contemporaneous pricing for the cost segregation study.
- Upon project completion, review and compare the as-built documents to the original contract documents for changes that affect the assets.

The Construction Team: General Contractor & Subcontractors

General contractors and subcontractors have the important distinction of being the best source of contemporaneous cost documentation. All project costs included within the contract will pass through the general contractor's books. Working with general contractors and subcontractors who understand cost segregation principles at the start and who have prior experience with buildings that have undergone a cost segregation study provides a clear advantage. Architects negotiating with owners for design services will often suggest "teams" or affiliated contractors and consultants, with whom they have a working relationship or who possess the special expertise needed to support the cost segregation process. There is clearly an opportunity for contractors to familiarize themselves with owners' needs in this area, and to market their ability to document project costs to support cost segregation studies.

Most contractors consider cost information relating to a stipulated sum contract to be proprietary. While this is clearly true, it is not the intent of the cost segregation study to decipher how much profit was made on a particular project. It is even less likely this would be possible without 100% of the job costs for the project. The recently-released IRS publication, *Cost Segregation Audit Techniques Guide* defines "cost" as the price the *owner paid* for the work or asset, not the cost incurred by the contractor. Overhead and profit added to the actual costs incurred by the contractor are considered acceptable, provided that the markups are within reasonable margins for the trade involved and can be substantiated, if necessary.

For most construction contracts, acceptable contemporaneous documentation would be considered the line item within an AIA G702 and G703 Application for Payment. This is recognized as the cost incurred by the owner, since this is what the owner pays for the work defined within the line item. Most contractors are not adverse to providing a detailed breakdown in their Applications for Payment, since this detail often has the benefit of preventing disputes regarding the progress of the work and the payment due.

General contractors and major subcontractors constructing a building that will undergo a cost segregation study should:

- Meet with the architect and cost segregation practitioner to review cost segregation (e.g., documentation) requirements. Compiling and organizing cost data for a study is time-consuming. Contractors should be aware of the expectations of the parties involved so that additional administrative fees can be included within the bid or fee.
- Review the bid (contract) documents carefully for contractual requirements relating to the cost segregation study.
- Ensure that the assets listed on the documents for recovery are clearly identified. Coordinate the plans with the specifications for confirmation.
- Estimate and construct assets as shown in the contract documents. For assets that have been identified for recovery, the old adage that "means and methods are the domain of the contractor" may not apply.
- Whenever possible, identify all parts, pieces, or components that comprise the asset when pricing it. Clearly identify these as part of the line item breakdown of the application for payment.
- Make sure change orders provide written description and details. Effects of a change order on a specific asset should be identified.
- Ensure that any changes to particular assets that occur during the course of construction are identified on the as-built drawings.
- Provide photo documentation of the project progress from mobilization to completion. Specific assets identified for recovery should be photographed in detail.
- Review contemporaneous cost documentation with the cost segregation practitioner to ensure that the documentation is sufficient.

(Subcontractors are included in the above list of items because, on many projects, their portion of the work is large enough to warrant their involvement at an early phase, just as with the general contractor.)

Conclusion

Owners considering applying cost segregation to new building projects have much to gain by assembling a cost segregation team early in the design process. Each team member must understand the responsibilities of his or her particular discipline in collecting and assembling the cost data that will act as the contemporaneous documentation for the cost segregation study. Alternative design features should be explored to better isolate assets that might otherwise be challenged.

Cost segregation practitioners who provide owners with a thorough presentation that highlights the benefits of a study can, in most cases, win the interest of even the most skeptical CEO or CFO. It is a service that often sells itself. Architects, engineers, and contractors should attend to the opportunity cost segregation presents to them as well. By learning about cost segregation and the requirements of a professional study, design and construction professionals can define and offer an array of support services that anticipate an owner's needs and add value in negotiation for projects.

The cost segregation concept is experiencing explosive growth, since it generates great benefits to owners. Under the current IRS regulations, a quality, professional cost segregation study will be of increasing importance, as owners at all levels become more educated on the concept of accelerated depreciation and the benefits of including cost segregation practitioners in the project development process.

Appendix A

Cost Segregation Audit Techniques Guide Excerpts

The following excerpts are from the IRS's *Cost Segregation Audit Techniques Guide*, made available to the public in 2004 and available online (at http://www.irs.gov/pub/irs-utl/costsegatg.htm). Please consult the IRS for full coverage.

Introduction

There are no standards for cost segregation studies. Thus, examiners will encounter a wide variety of studies and reports, as well as documentation. For example, some studies will be very brief. Other studies may be quite voluminous and complex. Regardless of the length of a study or the methodology used, a cost segregation study and report should always:

1. *classify* assets into property classes (e.g., land, land improvements, building, equipment, furniture and fixtures);
2. *explain the rationale* (including legal citations) for classifying assets as either § 1245 or § 1250 property; and,
3. *substantiate* the cost basis of each asset and reconcile total allocated costs to total actual costs.

What is a "Quality" Cost Segregation Study?

A "quality" cost segregation study is a study that is both accurate and well-documented with regard to the three points above. Quality studies greatly expedite the Service's review, thereby minimizing audit burden on all parties. A quality study contains a number of characteristics, which are set forth below.

Principal Elements of a Quality Cost Segregation Study

The 13 principal elements of a quality study are:

1. Preparation by an individual with expertise and experience
2. Detailed description of the methodology
3. Use of appropriate documentation
4. Interviews conducted with appropriate parties
5. Use of a common nomenclature
6. Use of a standard numbering system
7. Explanation of the legal analysis
8. Determination of unit costs and engineering "takeoffs"
9. Organization of assets into lists or groups
10. Reconciliation of total allocated costs to total actual costs
11. Explanation of the treatment of indirect costs
12. Identification and listing of Section 1245 property
13. Consideration of related aspects (e.g., IRC § 263A, Change in Accounting Method and Sampling Techniques)

Preparation by an Individual with Expertise and Experience

The preparation of cost segregation studies requires knowledge of both the construction process and the tax law involving property classifications for depreciation purposes. Unfortunately, there are no prescribed qualifications for cost segregation preparers. However, a preparer's credentials and level of expertise may have a bearing on the overall accuracy and quality of a study.

In general, a study by a construction engineer is more reliable than one conducted by someone with no engineering or construction background. However, the possession of specific construction knowledge is not the only criterion. Experience in cost estimating and allocation, as well as knowledge of the applicable law, are other important criteria.

A quality study identifies the preparer and always references his/her credentials, experience and expertise in the cost segregation area.

Detailed Description of the Methodology

While there are several common methodologies used in preparing a cost segregation study, the study may be based upon a variation or combination of methods.

A quality study always describes the methodology that was used and details the steps that were taken to classify assets and determine costs.

Use of Appropriate Documentation

A quality study uses contemporaneous documentation to classify assets and determine costs. Documentation supporting a quality study will vary, depending on whether a property is new or used or whether original construction documents are available. The documentation in a quality study for both new and used properties is detailed in the following sections.

New Construction

- Allocation of Land and Land Development Costs: A quality study explains the treatment of land and land development costs, (e.g., roads, sewer lines, storm drains, utility mains, survey and subdivision costs). Generally, these costs are allocated to non-depreciable land accounts. Also included in this account are the costs of improvements or land that are transferred to a local municipality (to obtain approval for subdividing or for a change in use).
- Site Visit: A quality study includes a site visit to gain a better perspective and understanding of the design and purpose of the project, as well as the use of specific assets. Before-and-after photographs are used to establish land and site preparation costs (i.e., surveying, clearing, grubbing, general grading and compaction).
- Blueprints, Construction Drawings, and Contract Payments: A quality study reviews all pertinent construction documentation. The taxpayer's capital expenditure request is reviewed to ascertain the intended functional use of a building and other assets included in the project. Site, architectural, and engineering plans, as well as "as-built" drawings, blueprints, and bid documents, are all reviewed and referenced in a quality study. The specific assets deemed to be § 1245 property are clearly highlighted or otherwise identified on the "as-built" drawings. Project specifications are analyzed to determine conformity to the blueprints. Purchase and change orders are also reviewed to ascertain cost information, changes in costs, and details of the work performed.
- A quality study reviews The General Contractor's Applications for Payment (AIA Forms G702 and G704) to ascertain what was actually paid for during construction. In addition, subcontractor payment applications, as well as invoices paid for items outside the scope of the general contractor's work, will be reviewed to provide greater insight and detail of the construction. Actual or estimated costs are cross-referenced to the supporting documentation.

Acquired or Used Properties

Unlike cost segregation studies performed on newly-constructed property, those performed on acquired or used property will likely be based on estimates and/or reconstructed costs.

- Purchase Price Allocations: A quality study documents how purchase price was allocated between land, land improvements, building and other assets. Land value is always determined first and is based on "highest and best use." In simple terms, highest and best use means the probable use of land that results in its highest value. The balance of the purchase price is then allocated to the building and to other assets.
- Addresses Physical Deterioration and Functional Obsolescence: The lack of cost records and the age of a property add to the uncertainty in determining its value or cost. In making this determination, a quality study always accounts for the physical deterioration and functional obsolescence of assets. It also provides the documents and the corroborating evidence used to determine values or costs.
- Site Visit: As for new construction, a quality study includes a site visit, as well as photographic evidence, to assist in identifying the assets and indetermining the allocations of values or costs.
- Review of Purchase or Lease Agreements and Appraisals: When original construction documents are not available, a quality study will support its allocations by using other corroborating evidence (e.g., purchase/lease agreement, appraisals).

A quality study will review the purchase agreement as a first step. If the property is leased, the lease agreement will be reviewed and documented. A quality study will also review any appraisals, if applicable. The availability of historical construction records will also be addressed in a quality study (i.e., if these are not available, the study will indicate what efforts were made to obtain these records).

Interviews Conducted with Appropriate Parties

Interviews with contractors and subcontractors, as well as with taxpayers and property managers, are quite valuable in ascertaining the specific use of a property and the construction process involved. A quality study documents all interviews conducted with appropriate parties, thus adding credibility to the depth and accuracy of its study. However, the examiner should recognize that subcontractor work details can be difficult to obtain since taxpayers generally have had no direct contact with them. In addition, general contractors may also be reluctant to share certain information because of confidentiality (e.g., profit margins).

Use of a Common Nomenclature

The use of creative or misleading nomenclature to describe property items, rather than common and clearly understood terms, detracts from the quality of a study. "Creative" descriptions may be used to disguise the true nature or character of an asset (e.g., a building sewage or water piping system referred to as "process piping"; an emergency exit sign termed "decorative placard").

A quality study always uses terminology consistent with the blueprints and other project documents (e.g., contract specifications, pay requests, etc.). The use of common and clearly understood terms facilitates Service review and avoids the confusion caused by misleading terms.

Use of a Standard Numbering System

The use of a standard numbering system, such as the Construction Specification Institute (CSI) MasterFormat Division, is helpful but not mandatory. A quality study numbers assets consistent with the contract bid documents and pay requests. This numbering system facilitates classifying property for computing depreciation and thus expedites the Service's exam.

The CSI format categorizes costs by specific building systems or components, such as concrete, carpentry, metals, woods and plastics, mechanical, electrical, and lighting. Other typical groupings of assets may include land, land improvements, furniture and fixtures, electrical systems, plumbing systems, equipment, etc.

Explanation of the Legal Analysis

A quality study contains a thorough legal analysis, including relevant citations, to support its § 1245 property classifications. While the treatment of some items may be fairly clear based on consistent judicial decisions, there are many instances in which court decisions may appear to be contradictory or to which the Service has not acquiesced. These apparent contradictions generally reflect the intensely factual basis that underlies the proper classification of property. As might be expected, the proper classification of property is the source of much audit controversy.

The legal discussion in a quality study recognizes these contradictions and attempts to reconcile them to the specific facts and circumstances of the property at issue. An accurate analysis of the statutes and judicial precedent adds to the overall quality of a study and facilitates the Service's review.

Determination of Unit Costs and Engineering "Take-Offs"

Once property items or assets have been identified and assigned to property classes (e.g., building and personal property), their respective costs must be determined. In order to determine a cost for each unit or class of property in a project or component system, total project costs (or total component system costs) must generally be broken down. This breakdown process is commonly known as engineering "take-offs".

In a quality study, engineering "take-offs" are carefully documented to show derived unit costs, and individual property units are clearly identified or highlighted on the "as-built" blueprints. For new construction, the cost of property items in an engineering take-off can generally be obtained from actual cost records. However, when actual costs are not available, costs must be estimated.

Cost estimates can vary widely depending on which estimating guide is used and whether costs are for "high" or "low" quality construction. In a quality study, cost estimates are always reconciled to an acquisition price, a total project cost, or to a component system cost to ensure the accuracy of an allocation. The proper use of an estimation technique is another frequent source of audit controversy. A quality study minimizes this controversy by clearly explaining and documenting the methodology used to assign costs to each asset.

Organization of Assets into Lists or Groups

Typically, a study lists assets by recovery period (e.g., land, land improvements, furniture and fixtures, electrical systems, plumbing systems, equipment). A quality study's asset listings tie to a taxpayer's fixed asset ledger, which also facilitates the Service's review.

Reconciliation of Total Allocated Costs to Total Actual Costs

It is important that the same estimating technique be used on all of the items that reconcile to a purchase price, a project cost, or to a particular component cost. If different methods or cost guides are used on different property items (e.g., one method for tangible personal property and a different method for the building), cost distortions arise. A quality study always reconciles total allocated costs to total actual costs in order to ensure the accuracy of its allocations.

A quality study also considers and lists separately-acquired § 1245 property to prevent possible duplication. For example, if the total project cost includes furniture, fixtures and equipment (FFE), then it is appropriate to allocate costs to those items. However, if FFE is acquired separately and not included in the total project cost, then it is not appropriate to assign costs to FFE.

Explanation of the Treatment of Indirect Costs

A quality study lists all the costs associated with a particular project, including both direct and indirect costs, and explains the treatment of any indirect costs. Direct costs are the labor and material costs for specific items or assets. Indirect costs, also referred to as "allocables," are intangible costs that are incident to the construction of a facility. Indirect costs must be allocated proportionately to the basis of the specific assets to which they relate.

Indirect costs also include expenditures that should not be allocated to the entire project but rather assigned to the property class to which they relate. Costs to survey and subdivide land, grade the land to prepare a building pad, and construct offsite improvements are generally allocable only to land. On the other hand, costs for building permits, general conditions, and contractor overhead and profit are typically allocated to assets on a pro-rata basis.

Generally, indirect costs do not relate to the placement of business machinery or furniture and fixtures since these assets are typically purchased and installed under separate contracts. However, indirect costs that specifically relate to components of personal property may be assigned to § 1245 property. For example, costs for special consultants (e.g., for computer wiring and process engineering) or costs to design the computer system may be assigned directly to that system. In addition, it may be reasonable to allocate certain indirect costs, such as liability insurance, bonds, and overhead/profit, where it can be shown that the total amount of the indirect costs is based upon the pro-rata cost of each class of property.

The treatment of indirect costs is another area of frequent controversy. A quality study explains the purpose of each indirect cost, describes its allocation, and explains any deviations from commonly accepted practice.

Identification and Listing of Section 1245 Property

A quality study lists § 1245 property (including amounts) and shows any § 1250 property reclassified to § 1245 property.

Consideration of Related Aspects (i.e., IRC § 263A, Change in Accounting Method and Sampling Techniques)

A quality study addresses related aspects, such as IRC § 263A, change in accounting method, and sampling techniques.

The uniform capitalization (UNICAP) rules of § 263A(a) require the capitalization of all direct costs and certain indirect costs allocable to real property and tangible personal property produced by the taxpayer. Self-constructed assets and property built under contract are treated as property "produced" by the taxpayer. Furthermore, § 263A(f) requires the capitalization of certain interest expense incurred in connection with the production of property.

Although the courts have not uniformly agreed, it is the position of the Service that a change in depreciation method, recovery period, or convention for depreciable property constitutes a change in accounting method. Therefore, the use of a cost segregation study to reclassify property and/or reallocate costs requires the consent of the Commissioner. Studies may utilize sampling techniques when taxpayers have a large number of substantially similar properties, such as retail or food stores. Studies may use such techniques as statistical sampling, modeling, or judgmental sampling.

When conducted properly, statistical sampling can be a reliable technique. However, improper sampling techniques may result in a final answer that does not accurately reflect a valid estimate. Factors addressed in a quality study's sampling technique include the definition of the population being sampled, the size of the population, a description of any stratification techniques, and the consideration of sampling error.

A modeling approach may also be used to segregate property costs. This approach uses created models to approximate the different types of units involved. If the models are properly analyzed, then this method may be reasonably accurate when applied to the entire population. However, the delineation of strata may be difficult and is often an area of controversy. Furthermore, issues may arise as to whether the sampling method is statistically valid.

Some studies may rely solely on a judgmental sampling technique, which carries a higher level of risk due to the elements of subjectivity involved. A judgment sample is typically selected on the basis of perceived similarities and is not statistically valid. However, under certain, limited circumstances, the use of a judgment sample may be appropriate. In such a case, the underlying basis for the selection of particular units in a judgment sample must be rational and supported by adequate data.

A quality study addresses these related audit issues and comments on the treatment of these items for tax purposes, especially where the amounts are restated for prior tax years.

Principal Elements of a Quality Cost Segregation Report

A cost segregation report reflects a study's methodology and conclusions. The amount of detail included in a report varies considerably since there is no standard or prescribed format. The following elements are found in a quality report.

1. Summary Letter/Executive Summary: A quality report contains a summary to identify: the preparer, the date of the study, the taxpayer (or client), the subject property, and the property components classified as land, land improvements, building, or personal property.
2. Narrative Report: A quality report discusses the theory, definitions, and rationale behind the study in the narrative section. This section generally includes a more detailed description of the property/facility (i.e., a physical description and an explanation of the use for which it is intended, as well as a legal description of the property and its location). In addition, the narrative section highlights the regulations, rulings, and court cases that support classifying assets as § 1245 property. The narrative also discusses the types and sources of data used (e.g., cost records, contracts, purchase agreements, published estimates) as well as how they were used.
3. Schedule of Assets: A quality report has a schedule of assets that are the focus of the study. Generally, this schedule ties directly to the taxpayer's depreciation records. When a taxpayer reallocates costs of assets already "on the books," a quality report clearly identifies the specific assets impacted (and includes depreciation records from both before and after the reallocation).
4. Schedule of Direct and Indirect Costs: A quality report lists all direct and indirect costs associated with a project. Indirect costs allocated to § 1245 property are clearly identified and explained. Separately-acquired assets are listed and discussed in the report to avoid duplication errors. Costs subject to IRC § 263A are also addressed.
5. Schedule of Property Units and Costs: A quality report provides a schedule of property units and costs (with property descriptions) that are segregated into land, § 1245 property, and § 1250 property. This schedule is the final product of the study and serves as the basis for computing depreciation.

6. Engineering Procedures: A quality report describes the engineering procedures and methodology for determining the cost of each property unit. It also identifies the specific taxpayer records that were reviewed and discusses whether actual cost records or estimating techniques were utilized to break costs into smaller components. A record of inspections and/or interviews is included as well. The use of a common nomenclature or a standard numbering system is also referenced and/or explained.
7. Statement of Assumptions and Limiting Conditions: A quality report describes the general understanding and conditions applicable to the report. This information may also provide an indication of the overall quality of the study.
8. Certification: A quality report certifies that the person who signed the report actually developed the analysis, opinions, and conclusions of the report. This section may also include the resumé or state the credentials and/or level of experience of the preparer.
9. Exhibits: A quality report generally includes various exhibits, such as the "Client Cost Sources" and the "Cost Source Reconciliation." These exhibits show the "book" (accounting) records on which the preparer relied in deriving total costs, and may include a reconciliation of the study to the fixed asset ledger. Photographs and/or videos may also be included as exhibits to assist in understanding the assets in the study.

Conclusions

The degree to which a cost segregation study/report conforms to these elements will likely dictate the scope and depth of an examination. A quality study and report will expedite the exam process and, ultimately, minimize audit burden on taxpayers, practitioners, and examiners alike.

Appendix B
Cost Segregation Issues in the Restaurant Industry

The following is a U.S. Treasury Department memorandum, dated December 8, 2003, outlining cost segregation issues specific to the restaurant industry.

Introduction

This memorandum is intended to provide direction to effectively utilize resources in the classification and examination of a taxpayer who is recovering costs through depreciation of tangible property used in the operation of a restaurant business. This Directive is not an official pronouncement of the law or the position of the Service and cannot be used, cited or relied upon as such.

Background

The crux of cost segregation is determining whether an asset is I.R.C. Section 1245 property (shorter cost recovery period property, 5 or 7 years) or Section 1250 property (longer cost recovery period property, 39, 31.5 or 15 years). The most common example of section 1245 property is depreciable personal property, such as equipment. The most common examples of section 1250 property are buildings and building components, which generally are not section 1245 property.[1]

[1] I.R.C. Section 1245 can apply to certain qualified recovery nonresidential real estate placed in service after 1980 and before 1987. *See* I.R.C. Section 1245 (a)(5).

The difference in recovery periods has placed the Internal Revenue Service and taxpayers in adversarial positions in determining whether an asset is section 1245 or 1250 property. Frequently, this causes the excessive expenditure of examination resources. The Director for the Retailers, Food, Pharmaceuticals and Healthcare Industry chartered a working group to address the most efficient way to approach cost segregation issues specific to the restaurant industry. The group produced the attached matrix and related definitions as a tool to reduce unnecessary disputes and foster consistent audit treatment.

Planning and Examination Guidance

Attached Exhibit A is a matrix recommending the categorization and general depreciation system recovery period of various restaurant assets. (For recovery periods under IRC section 168(g) alternative depreciation system see Revenue Procedure 87-56, 1987-2 CB 674.) If the taxpayer's tax return position for these assets is consistent with the recommendations in Exhibit A, examiners should not make adjustments to categorization and lives. If the taxpayer reports assets differently, then adjustments should be considered. The Industry intends to update Exhibit A regularly.

See also Revenue Procedure 2002-12, I.R.B. 2002-3, 374 (Jan. 07, 2002), for the proper treatment of smallwares.

LMSB DIRECTIVE ON COST SEGREGATION IN THE RESTAURANT INDUSTRY
EXHIBIT A

Asset	Description	Property Type	Recovery Period
Beverage Equipment	Equipment for storage and preparation of beverages and beverage delivery systems. Beverage equipment includes the refrigerators, coolers, dispensing systems, and the dedicated electrical, tubing or piping for beverage equipment. The dispensing system may be gravity, pump or gas driven.	1245	57.0 Distributive Trades and Services -- 5 Years
Canopies and Awnings	Readily removable equipment or apparatus used for providing shade or cover. Includes canopies that are largely decorative. Does not include canopies that are an integral part of a building's structural shell, such as in the casino industry.	1245	57.0 Distributive Trades and Services -- 5 Years
Ceiling	Includes all interior ceilings regardless of finish or décor, e.g. drywall or plaster ceilings, acoustic ceilings, suspended ceilings (including all hangers, frames, grids and tiles or panels), decorative metal or tin finishes, kitchen plastic panels, decorative panels, etc.	1250	Building or Building Component – 39 Years
Computers	Desktop and laptop computers, monitors, printers, and other peripheral equipment. Excludes Point of Sale (POS) systems and computers that are an integral part of other equipment (e.g., fire detection systems, heating or cooling systems, etc.).	1245	00.12 Information Systems – 5 Years
Concrete Foundations and Footings	Includes formwork, reinforcement, concrete block, and precast or cast-in-place work related to foundations and footings necessary for the proper setting of the building.	1250	Building or Building Component – 39 Years
	Foundations or footings for signs, light poles, canopies and other land improvements.	1250	00.3 Land Improvements – 15 Years
Copy Machines	Includes copiers and other duplicating equipment, calculators and adding machines. Excludes computers and computer peripheral equipment.	1245	00.13 Data Handling Equipment, except Computers – 5 Years
Doors - Eliason	Eliason Doors are specialty kitchen doors providing easy access to and from the kitchen area to the dining area. (Usually swings in both directions.)	1245	57.0 Distributive Trades and Services -- 5 Years
Doors and Windows	Includes interior and exterior doors and their hardware (such as doorknobs, closers, kickplates, hinges, locks, etc.) regardless of decoration. Exterior windows, including drive-through service windows and carousel windows and exterior storefront and vestibule.	1250	Building or Building Component – 39 Years

LMSB DIRECTIVE ON COST SEGREGATION IN THE RESTAURANT INDUSTRY
EXHIBIT A

Asset	Description	Property Type	Recovery Period
Doors and Windows - Accessories	Window accessories such as drapes, curtains, louver, post-construction tinting that is readily removable and interior decorative theme décor.	1245	57.0 Distributive Trades and Services -- 5 Years
Drive-Through Equipment	Drive-through equipment includes the order taking, food delivery and payment processing systems whether mechanical or electronic. This category does not include building elements such as doors, bays, or windows; see **Walls – Exterior**, and **Doors and Windows** for drive-through bays and windows.	1245	57.0 Distributive Trades and Services -- 5 Years
Electrical	Includes all components of the building electrical system necessary to provide general building services such as lighting, heating, ventilation, air conditioning and general electrical wiring and outlets used in the operation or maintenance of the building. See also **Kitchen Equipment Hook-up.**	1250	Building or Building Component – 39 Years
Electrical – Light Fixtures – Interior	Includes lighting such as recessed and lay-in lighting, night lighting, exit lighting, as well as decorative lighting fixtures that are the only artificial illumination in the building and along building walkways. See **Fire Protection Systems** for emergency and exit lighting.	1250	Building or Building Component – 39 Years
	Decorative light fixtures are light fixtures that are decorative in nature and not necessary for the operation of the building. In other words, if all the decorative lighting were turned off, the other sources of lighting would provide sufficient light for the building. If the decorative lighting is the only source of lighting, then it will be treated as section 1250 property.	1245	57.0 Distributive Trades and Services -- 5 Years
Electrical – Light Fixtures – Exterior	Exterior lighting whether decorative or not is considered section 1250 property to the extent that the lighting relates to the operation or maintenance of the building. This category includes building mounted lighting to illuminate walkways, entrances, parking, etc.	1250	Building or Building Component – 39 Years
	Plant grow lights or lighting that highlights <u>only</u> the landscaping or building exterior (but not parking areas or walkways) does not relate to the operation or maintenance of the building.	1245	57.0 Distributive Trades and Services -- 5 Years
	Pole mounted or freestanding outdoor lighting system to illuminate sidewalks, parking or recreation areas.	1250	00.3 Land Improvements – 15 Years

LMSB DIRECTIVE ON COST SEGREGATION IN THE RESTAURANT INDUSTRY
EXHIBIT A

Asset	Description	Property Type	Recovery Period
Elevators and Escalators	Elevators and escalators are permanently affixed to the building and intended to remain in place. They relate to the operation or maintenance of the building, and are therefore considered structural components.	1250	Building or Building Component – 39 Years
Equipment Installation	Expenses incurred in the installation of furnishings and restaurant equipment. Some examples include booths, tables, counters and interior theme décor.	1245	57.0 Distributive Trades and Services -- 5 Years
Fire Protection Equipment	Includes special fire detection or suppression systems located in equipment hoods or directly associated with a piece of equipment, and fire extinguishers for the protection against a particular hazard.	1245	57.0 Distributive Trades and Services -- 5 Years
Fire Protection Systems	Includes sprinkler heads, piping or plumbing, pumps, visual and audible alarms, alarm control panels, heat and smoke detection devices, fire escapes, fire doors, emergency exit lighting and signage, and wall mounted fire extinguishers necessary for the protection of the building.	1250	Building or Building Component – 39 Years
Fireplace	Includes masonry and gas fireplaces, flues, chimneys and other components of built-in fireplaces.	1250	Building or Building Component – 39 Years
Floor Covering - Carpet	For purposes of the restaurant industry, all carpeting will be treated as not permanently attached and not intended to be permanent. This category does not include rugs or tapestries that are considered artwork and do not suffer wear and tear (e.g. Persian rugs that may appreciate in value are considered artwork).	1245	57.0 Distributive Trades and Services -- 5 Years
Floor Covering - Permanent	Examples include ceramic or quarry tile, paving brick and other coverings cemented, mudded, or grouted to the floor; epoxy or sealers; wood flooring affixed with permanent adhesive or nailed or screwed in place and vinyl flooring affixed with permanent adhesive.	1250	Building or Building Component – 39 Years
Floors	Includes concrete slabs, and other floor systems. Floors include special treatments applied to or otherwise a permanent part of the floor. For example "superflat" finish, sloped drainage basins or raised perimeter and serving line curb or cooler, freezer and garbage room floors.	1250	Building or Building Component – 39 Years
Food Storage and Preparation Equipment	Food storage, cleaning, preparation, and delivery systems including all machinery, equipment, furniture and fixtures used to process food items from storage through delivery to the customer.	1245	57.0 Distributive Trades and Services -- 5 Years

LMSB DIRECTIVE ON COST SEGREGATION IN THE RESTAURANT INDUSTRY
EXHIBIT A

Asset	Description	Property Type	Recovery Period
Furnishings	Booths, tables, chairs, lockers, benches and other furniture needed in the business operation that is not a building component. See also **Office Furnishings.**	1245	57.0 Distributive Trades and Services -- 5 Years
Heating Ventilating and Air Conditioning (HVAC)	Includes all components of a central heating, ventilating and air conditioning system not specifically identified elsewhere.	1250	Building or Building Component – 39 Years
HVAC - Kitchen	Allocation of HVAC is not appropriate. Only separate Kitchen HVAC units that meet the sole justification test are included (i.e., machinery the sole justification for the installation of which is the fact that such machinery is required to meet temperature or humidity requirements which are essential for the operation of other machinery or the processing of materials or foodstuffs.) Kitchen HVAC may meet the sole justification test even though it incidentally provides for the comfort of employees, or serves, to an insubstantial degree, areas where such temperature or humidity requirements are not essential.	1245	57.0 Distributive Trades and Services -- 5 Years
Kitchen Equipment Hook-up	The kitchen equipment hook-ups include electrical, water, gas, sanitary plumbing, and ventilation systems for the kitchen equipment. Items in this category include wiring and outlets, plumbing, drainage and a hot water heater only in the kitchen area. The hook-up costs include all items used directly with a specific item of equipment and necessary to connect the equipment to the service source. For example, a dishwasher requires electric and plumbing hook-ups. Costs associated with connecting the dishwasher would only include the costs of wiring and plumbing from the dishwasher to the source of electricity (such as an outlet or junction box), plumbing, and drainage within the kitchen room. Use of reasonable functional allocation methods to determine the electrical and plumbing components directly related to the equipment is acceptable. This category does not include outlets in the dining area or kitchen hand sink plumbing; see **Electrical** and **Plumbing** categories.	1245	57.0 Distributive Trades and Services -- 5 Years

LMSB DIRECTIVE ON COST SEGREGATION IN THE RESTAURANT INDUSTRY
EXHIBIT A

Asset	Description	Property Type	Recovery Period
Millwork - Decorative	Decorative millwork is the decorative finish carpentry in the restaurant. Examples include detailed crown moldings, lattice work placed over finished walls or ceilings, cabinets and counters. The decorative millwork serves to enhance the overall theme of the restaurant and is not related to the operation of the building. Cabinets and counters in a restroom are excluded from this category; see **Restroom Accessories**.	1245	57.0 Distributive Trades and Services -- 5 Years
Millwork - General Building or Structural	General millwork is all building materials made of finished wood for example, doors and frames, window frames, sashes, porch work, mantels, panel work, stairways, and special woodwork. Includes pre-built wooden items brought to the site for installation and items constructed on site such as restroom cabinets, door jambs, moldings, trim, etc.	1250	Building or Building Component – 39 Years
Music and PA System	Equipment and apparatus used to provide amplified music or sound; also includes wiring. Does not include a public address system that is an integral part of a fire protection system.	1245	57.0 Distributive Trades and Services -- 5 Years
Office Furnishings	Includes desk, chair, credenza, file cabinet, table, or other furniture. Also includes telephone equipment. See **Furnishings** for tables, chairs, and other furniture in the dining area.	1245	00.11 Office Furniture, Fixtures, and Equipment – 7 Years
Plumbing	Includes all restroom plumbing fixtures (e.g. toilets) and piping; kitchen hand sinks; electric water coolers and all components of the building plumbing system not specifically identified elsewhere. Does not include water or gas connections directly to appliances or kitchen drainage and kitchen hot water heater; see **Kitchen Equipment Hook-up**.	1250	Building or Building Component – 39 Years
Point of Sale Equipment (POS)	Cash registers, computerized sales systems, and related peripheral equipment.	1245	57.0 Distributive Trades and Services -- 5 Years
Poles and Pylons	Light poles for parking areas and poles used in concrete footings for signage.	1250	00.3 Land Improvements – 15 Years
Restaurant Décor Accessories	Items in this category include decorative mobile props (interior or exterior). For example playground equipment, potted plants, hanging mirrors, ceiling fans, and theme related props (such as coat of arms, sporting equipment or memorabilia, artifacts, pictures, plaques, etc., provided these items are not non-depreciable artwork, antiques or collectibles).	1245	57.0 Distributive Trades and Services -- 5 Years

LMSB DIRECTIVE ON COST SEGREGATION IN THE RESTAURANT INDUSTRY
EXHIBIT A

Asset	Description	Property Type	Recovery Period
Restaurant Non-structural Theme Elements	Items in this category include interior non-load bearing decorative structures. These are items that do not function as part of the building and are not integrated with building elements such as wiring, plumbing or ventilation. For example a model castle constructed of gypsum board or plaster and wood studs would be considered a non-structural theme element that functions merely as ornamentation. On the other hand, a half wall whose function is to provide traffic control or space subdivision is not considered in this category, see **Walls - Interior Partitions**. Does not include decorative ceilings, see **Ceiling**.	1245	57.0 Distributive Trades and Services -- 5 Years
Restroom Accessories	Includes paper towel dispensers, electric hand dryers, towel racks or holders, cup dispensers, purse shelves, toilet paper holders, soap dispensers or holders, lotion dispensers, sanitary napkin dispensers and waste receptacles, coat hooks, grab bars, mirrors, shelves, vanity cabinets, counters and ashtrays and other items generally found in public restrooms that are built into or mounted on walls or partitions.	1250	Building or Building Component – 39 Years
Restroom Partitions	Includes shop made and standard manufacture toilet partitions; typically metal, but may be plastic or other materials.	1250	Building or Building Component – 39 Years
Roof	All elements of the roof including but not limited to joists, rafters, deck, shingles, vapor barrier, skylights, trusses, girders and gutters. Determination of whether decorative elements of a roof (e.g. false dormers, mansard) constitute structural building components depends on their integration with the overall roof not their load bearing capacity. If removal of the decorative element results in the direct exposure of building components to water, snow, wind, or moisture damage, or if the decorative element houses lighting fixtures, wiring, or other structural components, then the decorative elements are part of the overall roof system and are structural components of the building.	1250	Building or Building Component – 39 Years
Security Systems	Includes security equipment for the protection of the building (and its contents) from theft or vandalism and protection of employees from assault. Examples include window and door locks, card key access systems, keyless entry systems, motion detection and alarm systems, etc.	1250	Building or Building Component – 39 Years

LMSB DIRECTIVE ON COST SEGREGATION IN THE RESTAURANT INDUSTRY
EXHIBIT A

Asset	Description	Property Type	Recovery Period
Signs	Interior and Exterior Signs used for menu display, theme identity, and any signage not pertinent to the operation of the building. Does not include exit signs; see **Fire Protection Systems** for exit signs.	1245	57.0 Distributive Trades and Services -- 5 Years
Site Grading and Excavation	In general, land preparation costs are not considered to be tangible property subject to depreciation. The grading of land involves moving soil for the purpose of changing the ground surface. It produces a more level surface and generally provides an improvement that adds value to the land. Includes one time cost of demolition, clearing, fill or excavation to allow development of land.		Land
	Clearing, grading, excavating and removal costs directly associated with the construction of buildings and building components are part of the cost of construction of the building.	1250	Building or Building Component – 39 Years
	Clearing, grading, excavating and removal costs directly associated with the construction of sidewalks, parking areas, roadways and other depreciable land improvements are part of the cost of construction of the improvements.	1250	00.3 Land Improvements – 15 Years
Site Utilities	Site utilities are the systems that are used to distribute utility services from the property line to the restaurant building. Water and gas services are connected to the building by underground piping. Overhead or underground lines connect electric service. Not all restaurants will have such costs. For example, mall food court restaurants would not incur costs for site utilities since the lessor already has provided them.	1250	Building or Building Component – 39 Years
Site Work	Site work includes site drainage, sewers, roads, sidewalks, paving, curbing, general site improvements, landscaping not adjacent to the building, site fencing and enclosures, playground fencing, and all other site improvements not directly related to the building.	1250	00.3 Land Improvements – 15 Years
Stonework	This category includes exterior decorative stonework embedded in exterior half walls, such as patio half walls, that are an integral part of a building's structural shell. In the context of the restaurant industry, such half walls are related to the operation or maintenance of the building.	1250	Building or Building Component – 39 Years
	Includes patio stonework imbedded in the ground and applied to exterior half walls that are not an integral part of the building's structural shell.	1250	00.3 Land Improvements – 15 Years

LMSB DIRECTIVE ON COST SEGREGATION IN THE RESTAURANT INDUSTRY
EXHIBIT A

Asset	Description	Property Type	Recovery Period
Trash Enclosure	Enclosures for waste receptacles that are attached to the building. Typically constructed of the same materials as the building shell with either interior or exterior access. These trash enclosures are an integral part of the building shell and cannot be moved without damage to the underlying building.	1250	Building or Building Component – 39 Years
Upholstery	Any material used in the coverage and protection of furnishings.	1245	57.0 Distributive Trades and Services -- 5 Years
Wall Coverings - Permanent	Includes interior and exterior paint; ceramic or quarry tile, stone, brick and other finishes affixed with mortar, cement or grout; paneling, wainscoting and other wood finishes affixed with nails, screws or permanent adhesives; and sanitary kitchen wall panels such as fiberglass, stainless steel and plastic wall panels.	1250	Building or Building Component – 39 Years
Wall Coverings - Removable	Includes strippable wallpaper that causes no damage to the underlying wall or wall surface.	1245	57.0 Distributive Trades and Services -- 5 Years
Walls - Exterior Walls - Interior Partitions	Includes all exterior walls and building support regardless of construction materials. Exterior walls may include columns, posts, beams, girders, curtain walls, tilt up panels, studs, framing, sheetrock, insulation, windows, doors, exterior façade, brick, masonry, etc. Also includes drive-through bay, windows, and doors.	1250	Building or Building Component – 39 Years
	Includes all load bearing interior partitions regardless of construction. Also includes non-load bearing partitions regardless of height (typically constructed of studs and sheetrock or other materials) that divide or create rooms or provide traffic control where the partition cannot be 1) readily removed and remain in substantially the same condition after removal as before, or 2) moved and reused, stored or sold in its entirety. Includes rough carpentry and plaster, dry wall or gypsum board, and other finishes.	1250	Building or Building Component – 39 Years

Appendix C
Calculating the Surface Area of Cylinders, Pyramids, & Cones

Estimators often need to calculate the surface area of a three-dimensional shape, such as a cylinder, pyramid, or cone—for example, to determine the cost to paint one of these shapes. The outside surface area of a three-dimensional shape is referred to as its *lateral area*. The formula for calculating the lateral area of a cylinder is:

> **A_L = C x H, where C = circumference of the cylinder base and H = height of the cylinder.**

The lateral area of a pyramid can be expressed as:

> **A_L = p x 1/2 x h_s, where p= perimeter of the base of the pyramid where the base is a regular polygon and h_s = slant height of the pyramid.**

The slant height of the pyramid is definded as a line drawn from the *vertex*, or converging point at the top of the pyramid, to the center of any one side of the base.

The lateral area of a cone is the area of its tapering side. It can be expressed as:

> **A_L = C_b x 1/2 x h_s, where C_b = circumference of the base of the cone and h_s = slant height of the cone.**

The slant height of the cone is defined as a straight line drawn from the vertex of the cone to the circumference. It can be calculated by solving for the hypotenuse in the Pythagorean Theorem where the altitude of the cone is its height perpendicular to the base, and the radius of the base of the cone is its base.

Appendix D Using RSMeans Cost Data

Users of *Means Building Construction Cost Data* are chiefly interested in obtaining quick, reasonable, average prices for building construction items. This is the primary purpose of the annual book—to eliminate guesswork when pricing unknowns. Many people use the cost data, whether for bids or verification of quotations or budgets, without being fully aware of how the prices are obtained and derived. Without this knowledge, this vast resource is not being used to its fullest advantage. In addition to the basic cost data, the book also contains a wealth of information to aid the estimator, the contractor, the designer, and the owner to better plan and manage building construction projects. Productivity data is provided in order to assist with scheduling. National labor rates are analyzed. Tables and charts for location and time adjustments are included and help the estimator to tailor the prices to a specific location. And ultimately, the book is a list of line items for over 20,000 individual components of construction. This information, outlined in the CSI MasterFormat divisions, provides an invaluable checklist to the construction professional to assure that all required items are included in a project.

Format & Data

The major portion of *Means Building Construction Cost Data* is the Unit Price section. This is the primary source of unit cost data and is organized according to the 16 divisions of the MasterFormat. This index was developed by representatives of all parties concerned with the building construction industry and has been adopted by the American Institute of Architects (AIA), the Associated General Contractors of America, Inc. (AGC), and the Construction Specifications Institute, Inc. (CSI).

The MasterFormat Divisions:

Division 1 – General Requirements

Division 2 – Site Construction

Division 3 – Concrete

Division 4 – Masonry

Division 5 – Metals

Division 6 – Wood and Plastics

Division 7 – Thermal and Moisture Protection

Division 8 – Doors and Windows

Division 9 – Finishes

Division 10 – Specialties

Division 11 – Equipment

Division 12 – Furnishings

Division 13 – Special Construction

Division 14 – Conveying Systems

Division 15 – Mechanical

Division 16 – Electrical

In addition to the sixteen MasterFormat divisions of the Unit Price section, Division 17, Square Foot and Cubic Foot Costs, presents consolidated data from over ten thousand actual reported construction projects and provides information based on total project costs as well as costs for major components. Following Division 17 is the Reference Section, which provides supporting cost, design, and reference information for the Unit Price section. Individual materials and construction methods are presented in depth and detail in this section. Also included in this section are Change Orders, Crew Tables, City Cost Index, Abbreviations, and finally, the book index.

The prices as presented in *Means Building Construction Cost Data* are national averages. Material and equipment costs are developed through annual contact with manufacturers, dealers, distributors, and contractors throughout the United States. Means' staff of engineers is constantly updating prices and keeping abreast of changes and fluctuations within the industry. Labor rates are the national average of each trade as determined from union agreements from thirty major U.S. cities. Throughout the calendar year, as new wage agreements are negotiated, labor costs should be factored accordingly.

There are various factors and assumptions on which the costs, as presented in *Means Building Construction Cost Data*, have been based:

> *Quality* – The costs are based on methods, materials, and workmanship in accordance with U.S. Government standards and represent good, sound construction practice.
>
> *Overtime* – The costs as presented include *no* allowance for overtime. If overtime or premium time is anticipated, labor costs must be factored accordingly.
>
> *Productivity* – The daily output figures are based on an eight hour workday, during daylight hours.

Size of Project – Costs in *Means Building Construction Cost Data* are based on commercial and industrial buildings which cost $1,000,000 and up. Large residential projects are also included.

Local Factors – Weather conditions, season of the year, local union restrictions, and unusual building code requirements can all have a significant impact on construction costs. The availability of a skilled labor force, sufficient materials, and even adequate energy and utilities will also affect costs. These factors vary in impact and are not dependent upon location. They must be reviewed for each project in every area.

In the presentation of prices in *Means Building Construction Cost Data*, certain rounding rules are employed to make the numbers easy to use without significantly affecting accuracy. The rules are used consistently and are as follows:

Prices From	**Rounded to the nearest**
$0.01 to $5.00	$.01
$5.01 to $20.00	$.05
$20.01 to $100.00	$.50
$100.01 to $300.00	$1.00
$300.01 to $1,000.00	$5.00
$1,000.01 to $10,000.00	$25.00
$10,000.01 to $50,000.00	$100.00
$50,000.01 and above	$500.00

Unit Price Section

The Unit Price section of *Means Building Construction Cost Data* contains a great deal of information in addition to the unit cost for each construction component. Figure D.1 is a typical page, showing costs for brick veneer. Note that prices are included for more than ten types of brick and ten bonding patterns and course types. In addition, each type and bond is priced per thousand brick as well as per square foot of wall area. The information and cost data is broken down and itemized in this way to provide not only for the most detailed pricing possible, but also to accommodate different methods of quantity takeoff and estimating.

Within each individual line item, there is a description of the construction component, information regarding typical crews designated to perform the work, and the daily output of each crew. Costs are presented bare, or unburdened, as well as with markups for overhead and profit. Figure D.2 is a graphic representation of how to use the Unit Price section as presented in *Means Building Construction Cost Data.*

04200 | Masonry Units

	04210 | Clay Masonry Units			CREW	DAILY OUTPUT	LABOR-HOURS	UNIT	2002 BARE COSTS MAT.	LABOR	EQUIP.	TOTAL	TOTAL INCL O&P	
100	0010	**COMMON BUILDING BRICK** C62, TL lots, material only	R04210 -120										100
	0020	Standard, minimum					M	270			270	297	
	0050	Average (select)					"	315			315	345	
120	0010	**BRICK VENEER** Scaffolding not included, truck load lots	R04210 -120										120
	0015	Material costs incl. 3% brick and 25% mortar waste											
	0020	Standard, select common, 4" x 2-2/3" x 8" (6.75/S.F.)	R04210 -180	D-8	1.50	26.667	M	360	740		1,100	1,525	
	0050	Red, 4" x 2-2/3" x 8", running bond			1.50	26.667		400	740		1,140	1,575	
	0100	Full header every 6th course (7.88/S.F.)	R04210 -500		1.45	27.586		400	765		1,165	1,625	
	0150	English, full header every 2nd course (10.13/S.F.)			1.40	28.571		395	790		1,185	1,650	
	0200	Flemish, alternate header every course (9.00/S.F.)			1.40	28.571		395	790		1,185	1,650	
	0250	Flemish, alt. header every 6th course (7.13/S.F.)			1.45	27.586		400	765		1,165	1,625	
	0300	Full headers throughout (13.50/S.F.)			1.40	28.571		395	790		1,185	1,650	
	0350	Rowlock course (13.50/S.F.)			1.35	29.630		395	820		1,215	1,675	
	0400	Rowlock stretcher (4.50/S.F.)			1.40	28.571		400	790		1,190	1,675	
	0450	Soldier course (6.75/S.F.)			1.40	28.571		400	790		1,190	1,675	
	0500	Sailor course (4.50/S.F.)			1.30	30.769		400	850		1,250	1,750	
	0601	Buff or gray face, running bond, (6.75/S.F.)			1.50	26.667		400	740		1,140	1,575	
	0700	Glazed face, 4" x 2-2/3" x 8", running bond			1.40	28.571		1,500	790		2,290	2,875	
	0750	Full header every 6th course (7.88/S.F.)			1.35	29.630		1,475	820		2,295	2,850	
	1000	Jumbo, 6" x 4" x 12",(3.00/S.F.)			1.30	30.769		1,250	850		2,100	2,675	
	1051	Norman, 4" x 2-2/3" x 12" (4.50/S.F.)			1.45	27.586		735	765		1,500	1,975	
	1100	Norwegian, 4" x 3-1/5" x 12" (3.75/S.F.)		D-8	1.40	28.571	M	575	790		1,365	1,850	
	1150	Economy, 4" x 4" x 8" (4.50 per S.F.)			1.40	28.571		560	790		1,350	1,850	
	1201	Engineer, 4" x 3-1/5" x 8", (5.63/S.F.)			1.45	27.586		365	765		1,130	1,575	
	1251	Roman, 4" x 2" x 12", (6.00/S.F.)			1.50	26.667		765	740		1,505	1,975	
	1300	S.C.R. 6" x 2-2/3" x 12" (4.50/S.F.)			1.40	28.571		900	790		1,690	2,225	
	1350	Utility, 4" x 4" x 12" (3.00/S.F.)			1.35	29.630		1,075	820		1,895	2,450	
	1360	For less than truck load lots, add						10			10	11	
	1400	For battered walls, add							30%				
	1450	For corbels, add							75%				
	1500	For curved walls, add							30%				
	1550	For pits and trenches, deduct							20%				
	1999	Alternate method of figuring by square foot											
	2000	Standard, sel. common, 4" x 2-2/3" x 8", (6.75/S.F.)		D-8	230	.174	S.F.	2.69	4.82		7.51	10.35	
	2020	Standard, red, 4" x 2-2/3" x 8", running bond (6.75/SF)			220	.182		2.69	5.05		7.74	10.70	
	2050	Full header every 6th course (7.88/S.F.)			185	.216		3.13	6		9.13	12.65	
	2100	English, full header every 2nd course (10.13/S.F.)			140	.286		4.02	7.90		11.92	16.55	
	2150	Flemish, alternate header every course (9.00/S.F.)			150	.267		3.57	7.40		10.97	15.30	
	2200	Flemish, alt. header every 6th course (7.13/S.F.)			205	.195		2.84	5.40		8.24	11.40	
	2250	Full headers throughout (13.50/S.F.)			105	.381		5.35	10.55		15.90	22	
	2300	Rowlock course (13.50/S.F.)			100	.400		5.35	11.10		16.45	23	
	2350	Rowlock stretcher (4.50/S.F.)			310	.129		1.80	3.57		5.37	7.50	
	2400	Soldier course (6.75/S.F.)			200	.200		2.69	5.55		8.24	11.45	
	2450	Sailor course (4.50/S.F.)			290	.138		1.80	3.82		5.62	7.85	
	2600	Buff or gray face, running bond, (6.75/S.F.)			220	.182		2.85	5.05		7.90	10.90	
	2700	Glazed face brick, running bond			210	.190		9.90	5.30		15.20	19	
	2750	Full header every 6th course (7.88/S.F.)			170	.235		11.55	6.50		18.05	22.50	
	3000	Jumbo, 6" x 4" x 12" running bond (3.00/S.F.)			435	.092		3.53	2.55		6.08	7.80	
	3050	Norman, 4" x 2-2/3" x 12" running bond, (4.5/S.F.)			320	.125		3.50	3.46		6.96	9.15	
	3100	Norwegian, 4" x 3-1/5" x 12" (3.75/S.F.)			375	.107		2.11	2.95		5.06	6.85	
	3150	Economy, 4" x 4" x 8" (4.50/S.F.)			310	.129		2.50	3.57		6.07	8.25	
	3200	Engineer, 4" x 3-1/5" x 8" (5.63/S.F.)			260	.154		2.05	4.26		6.31	8.80	
	3250	Roman, 4" x 2" x 12" (6.00/S.F.)			250	.160		4.52	4.43		8.95	11.75	
	3300	SCR, 6" x 2-2/3" x 12" (4.50/S.F.)			310	.129		4.14	3.57		7.71	10.05	
	3350	Utility, 4" x 4" x 12" (3.00/S.F.)			450	.089		3.20	2.46		5.66	7.30	
	3400	For cavity wall construction, add							15%				
	3450	For stacked bond, add							10%				
	3500	For interior veneer construction, add							15%				
	3550	For curved walls, add							30%				

Figure D.1

Credit: Excerpted from *Unit Price Estimating Methods, Third Edition,* RSMeans

Line Numbers

Every construction item in the Means unit price cost data books has a unique line number. This line number acts as an address so that each item can be quickly located and/or referenced. The numbering system is based on the MasterFormat classification by division. In Figure D.2, note the bold number in reverse type, 03310. This number represents the MEANS subdivision, in this case, Structural Concrete, of the MasterFormat Division 3 – Concrete. All 16 MasterFormat divisions are organized in this manner. Within each subdivision, the data is broken down into major classifications. These major classifications are listed alphabetically and are designated by bold type for both numbers and descriptions. Each item, or line, is further defined by an individual number. As shown in Figure D.2, the full line number for each item consists of: MasterFormat Division – MEANS subdivision – a major classification number – an item line number. Each full line number describes a unique construction element. For example, in Figure D.1, the line number for Norman brick veneer, running bond (per square foot) is 04210-120-3050.

Line Description

Each line has a text description of the item for which costs are listed. The description may be self-contained and all inclusive. If indented, items are delineations (by size, color, material, etc.) or breakdowns of previously described items. An index is provided in the back of *Means Building Construction Cost Data* to aid in locating particular items.

Crew

For each construction element (each line item) a minimum typical crew is designated as appropriate to perform the work. The crew may include one or more trades, foremen, craftsmen and helpers, and any equipment required for proper installation of the described item. If an individual trade installs the item using only hand tools, the smallest efficient number of tradesmen will be indicated (1 Carp, 2 Carp, etc.). Abbreviations for trades are shown in Figure D.3. If more than one trade is required to install the item and/or if powered equipment is needed, a crew number will be designated (B-5, D-3, etc.) A complete listing of crews is presented following the Reference Section in the back of *Means Building Construction Cost Data*. Each crew breakdown contains the following components:

1. Number and type of workers designated.
2. Number, size, and type of any equipment required.
3. Hourly labor costs listed two ways: bare (base rate including fringe benefits) and including installing contractor's overhead and profit (billing rate). See Figure D.3.
4. Daily equipment costs, based on the weekly equipment rental cost divided by 5, plus the hourly operating cost, times 8 hours. This cost is listed two ways: as a bare cost and with a 10% markup to cover handling and management costs.
5. Labor and equipment are broken down further: cost per labor-hour for labor, and cost per labor-hour for the equipment.
6. The total daily labor-hours for the crew.
7. The total bare costs per day for the crew, including equipment.
8. The total daily cost of the crew including the installing contractor's overhead and profit.

How to Use the Unit Price Pages

The following is a detailed explanation of a sample entry in the Unit Price Section. Next to each bold number below is the item being described with appropriate component of the sample entry following in parenthesis. Some prices are listed as bare costs, others as costs that include overhead and profit of the installing contractor. In most cases, if the work is to be subcontracted, the general contractor will need to add an additional markup (R.S. Means suggests using 10%) to the figures in the column "Total Incl. O&P."

1 Division Number/Title (03300/Cast-In-Place Concrete)

Use the Unit Price Section Table of Contents to locate specific items. The sections are classified according to the CSI MasterFormat (1995 Edition).

2 Line Numbers (03310 240 3900)

Each unit price line item has been assigned a unique 12-digit code based on the CSI MasterFormat classification.

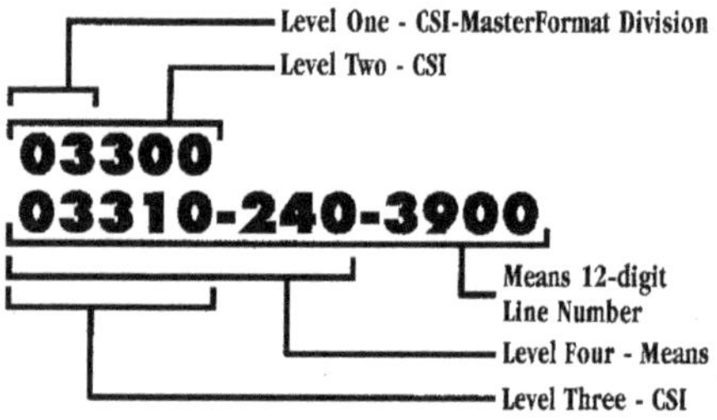

3 Description (Concrete-In-Place, etc.)

Each line item is described in detail. Sub-items and additional sizes are indented beneath the appropriate line items. The first line or two after the main item (in boldface) may contain descriptive information that pertains to all line items beneath this boldface listing.

Items which include the symbol **CN** are updated in the Key Material Price Section of *The Change Notice* quarterly publication.

4 Reference Number Information

R03310 -010 You'll see reference numbers shown in bold rectangles at the beginning of some sections. These refer to related items in the Reference Section, visually identified by a vertical gray bar on the edge of pages.

The relation may be: (1) an estimating procedure that should be read before estimating, (2) an alternate pricing method, or (3) technical information.

The "R" designates the Reference Section. The numbers refer to the MasterFormat classification system.

It is strongly recommended that you review all reference numbers that appear within the section in which you are working.

Note: Not all reference numbers appear in all Means publications.

03300 | Cast-In-Place Concrete

	03310 | Structural Concrete	CREW	DAILY OUTPUT	LABOR-HOURS	UNIT	2002 BARE COSTS MAT.	LABOR	EQUIP.	TOTAL	TOTAL INCL O&P	
240 0010	**CONCRETE IN PLACE** including forms (4 uses), reinforcing steel, including finishing unless otherwise indicated R03310 -010										240
0050											
3800	Footings, spread under 1 C.Y.	C-14C	38.07	2.942	C.Y.	100	84.50	.94	185.44	244	
3850	Over 5 C.Y.		81.04	1.382		92	40	.44	132.44	164	
3900	Footings, strip, 18" x 9", plain		41.04	2.729		91	78.50	.87	170.37	225	
3950	36" x 12", reinforced		61.55	1.820		92.50	52.50	.58	145.58	185	
4000	Foundation mat, under 10 C.Y.		38.67	2.896		124	83.50	.93	208.43	[illegible]	
4050	Over 20 C.Y.		[illegible]	[illegible]		110	57	.64	167.64	[illegible]	
4200	Grade walls, 8" thick, 8' high	C-14D	[illegible]	[illegible]		126	130	15.65	271.65	[illegible]	
4250	14' high		27.26	7.337		149	219	26.50	394.50	540	
4260	12" thick, 8' high		64.32	3.109		111	93	11.15	215.15	279	
4270	14' high		40.01	4.999		119	149	17.95	285.95	385	

Figure D.2

Credit: Excerpted from *Unit Price Estimating Methods, Third Edition,* RSMeans

		A		B	C	D	E	F	G	H	I
		Base Rate Incl. Fringes		Workers' Comp. Ins.	Average Fixed Overhead	Overhead	Profit	Total Overhead & Profit		Rate with O & P	
Abbr.	Trade	Hourly	Daily					%	Amount	Hourly	Daily
Skwk	Skilled Workers Average (35 trades)	$30.95	$247.60	16.8%	16.5%	13.0%	10%	56.3%	$17.40	$48.35	$386.80
	Helpers Average (5 trades)	22.75	182.00	18.5		11.0		56.0	12.75	35.50	284.00
	Foreman Average, Inside ($.50 over trade)	31.45	251.60	16.8		13.0		56.3	17.70	49.15	393.20
	Foreman Average, Outside ($2.00 over trade)	32.95	263.60	16.8		13.0		56.3	18.55	51.50	412.00
Clab	Common Building Laborers	23.45	187.60	18.1		11.0		55.6	13.05	36.50	292.00
Asbe	Asbestos/Insulation Workers/Pipe Coverers	33.45	267.60	16.2		16.0		58.7	19.65	53.10	424.80
Boil	Boilermakers	36.25	290.00	14.7		16.0		57.2	20.75	57.00	456.00
Bric	Bricklayers	30.50	244.00	16.0		11.0		53.5	16.30	46.80	374.40
Brhe	Bricklayer Helpers	23.50	188.00	16.0		11.0		53.5	12.55	36.05	288.40
Carp	Carpenters	30.00	240.00	18.1		11.0		55.6	16.70	46.70	373.60
Cefi	Cement Finishers	28.70	229.60	10.6		11.0		48.1	13.80	42.50	340.00
Elec	Electricians	35.45	283.60	6.7		16.0		49.2	17.45	52.90	423.20
Elev	Elevator Constructors	37.10	296.80	7.7		16.0		50.2	18.60	55.70	445.60
Eqhv	Equipment Operators, Crane or Shovel	32.35	258.80	10.6		14.0		51.1	16.55	48.90	391.20
Eqmd	Equipment Operators, Medium Equipment	31.20	249.60	10.6		14.0		51.1	15.95	47.15	377.20
Eqlt	Equipment Operators, Light Equipment	29.80	238.40	10.6		14.0		51.1	15.25	45.05	360.40
Eqol	Equipment Operators, Oilers	26.65	213.20	10.6		14.0		51.1	13.60	40.25	322.00
Eqmm	Equipment Operators, Master Mechanics	32.80	262.40	10.6		14.0		51.1	16.75	49.55	396.40
Glaz	Glaziers	30.00	240.00	13.8		11.0		51.3	15.40	45.40	363.20
Lath	Lathers	28.75	230.00	11.1		11.0		48.6	13.95	42.70	341.60
Marb	Marble Setters	30.10	240.80	16.0		11.0		53.5	16.10	46.20	369.60
Mill	Millwrights	31.75	254.00	10.6		11.0		48.1	15.25	47.00	376.00
Mstz	Mosaic & Terrazzo Workers	29.25	234.00	9.8		11.0		47.3	13.85	43.10	344.80
Pord	Painters, Ordinary	27.15	217.20	13.8		11.0		51.3	13.95	41.10	328.80
Psst	Painters, Structural Steel	27.90	223.20	48.4		11.0		85.9	23.95	51.85	414.80
Pape	Paper Hangers	27.10	216.80	13.8		11.0		51.3	13.90	41.00	328.00
Pile	Pile Drivers	29.80	238.40	24.9		16.0		67.4	20.10	49.90	399.20
Plas	Plasterers	28.10	224.80	15.8		11.0		53.3	15.00	43.10	344.80
Plah	Plasterer Helpers	23.70	189.60	15.8		11.0		53.3	12.65	36.35	290.80
Plum	Plumbers	35.95	287.60	8.3		16.0		50.8	18.25	54.20	433.60
Rodm	Rodmen (Reinforcing)	34.25	274.00	28.3		14.0		68.8	23.55	57.80	462.40
Rofc	Roofers, Composition	26.60	212.80	32.6		11.0		70.1	18.65	45.25	362.00
Rots	Roofers, Tile & Slate	26.75	214.00	32.6		11.0		70.1	18.75	45.50	364.00
Rohe	Roofers, Helpers (Composition)	19.80	158.40	32.6		11.0		70.1	13.90	33.70	269.60
Shee	Sheet Metal Workers	35.10	280.80	11.7		16.0		54.2	19.00	54.10	432.80
Spri	Sprinkler Installers	36.20	289.60	8.7		16.0		51.2	18.55	54.75	438.00
Stpi	Steamfitters or Pipefitters	36.20	289.60	8.3		16.0		50.8	18.40	54.60	436.80
Ston	Stone Masons	30.65	245.20	16.0		11.0		53.5	16.40	47.05	376.40
Sswk	Structural Steel Workers	34.25	274.00	39.8		14.0		80.3	27.50	61.75	494.00
Tilf	Tile Layers	29.15	233.20	9.8		11.0		47.3	13.80	42.95	343.60
Tilh	Tile Layers Helpers	23.35	186.80	9.8		11.0		47.3	11.05	34.40	275.20
Trlt	Truck Drivers, Light	24.30	194.40	14.9		11.0		52.4	12.75	37.05	296.40
Trhv	Truck Drivers, Heavy	25.00	200.00	14.9		11.0		52.4	13.10	38.10	304.80
Sswl	Welders, Structural Steel	34.25	274.00	39.8		14.0		80.3	27.50	61.75	494.00
Wrck	*Wrecking	23.45	187.60	41.2	▼	11.0	▼	78.7	18.45	41.90	335.20

*Not included in averages

Figure D.3

Credit: Excerpted from ***Unit Price Estimating Methods, Third Edition***, RSMeans

The total daily cost of the craftsmen involved or the total daily crew indicated is used to calculate the unit installation cost for each item (for both bare costs and cost including overhead and profit).

The crew designation does not mean that this is the only crew that can perform the work. Crew size and content have been developed and chosen based on practical experience and feedback from contractors and represent a labor and equipment makeup commonly found in the industry. The most appropriate crew for a given task is best determined based on particular project requirements. Unit costs may vary if crew sizes or content are significantly changed.

These figures represent costs as included in the bare daily crew costs and are calculated as follows:

Line number:	01590-200-1910
Equipment:	Grader, self-propelled, 30,000 lb.
Rent per Week:	\$1,470
Hourly Operating Cost:	\$17.20

$$\frac{\text{Weekly rental}}{\text{5 days per week}} + (\text{Hourly Oper. Cost} \times 8 \text{ hrs./day}) = \frac{\text{Crew's Equipment}}{\text{Cost per Day}}$$

$$\frac{\$1{,}470}{5} + (\$17.20 \times 8) = \$431.60$$

The Crew's Equipment Cost is basically the daily cost of equipment. This figure is based on rental by the week. Note how the equipment costs in the example above are used in developing the crew cost for Crew B-26 from *Means Building Construction Cost Data.* The daily equipment costs, including overhead and profit, contain a 10% fee added to the bare costs to cover the handling and overhead associated with use of the equipment.

The daily output represents the number of units that the designated minimum crew will install in one 8 hour day. (See "Units" section.) These figures have been determined from construction experience under actual working conditions. With a designation of the appropriate crew and determination of the daily output, the unit costs for installation are easily calculated:

$$\frac{\text{Daily Crew Cost (\$/day)}}{\text{Daily Output (units/day)}} = \text{Unit Installation Cost (\$/unit)}$$

The labor-hours per unit can also be easily determined using the data provided. A labor-hour is the equivalent of one worker working for one hour, and is a relative indicator of productivity. To calculate labor-hours per unit:

$$\frac{\text{Total Daily Crew Labor-hours}}{\text{Daily Output}} = \frac{\text{Labor-hours/Day}}{\text{Units/Day}} = \text{Labor-hours per Unit}$$

The resulting units of productivity can be very useful as an aid for scheduling. Manpower allocations are more readily determined and time durations can be calculated based on crew size.

Unit

The unit column (see Figures D.1 and D.2) defines the component for which the costs have been calculated. It is this "unit" on which Unit Price Estimating is based. The units, as used, represent standard estimating and quantity takeoff procedures. However, the estimator should always check to be sure that the units taken off are the same as those priced. Note in Figure D.1 that the same items are priced based on two very different units—per thousand brick and per square foot. A list of standard abbreviations is included at the back of *Means Building Construction Cost Data*.

Bare Costs

The four columns listed under "Bare Costs"—"Material," "Labor," "Equip.," and "Total"—represent the actual cost of construction items to the contractor. In other words, bare costs are those which *do not* include the overhead and profit of the installing contractor, whether a subcontractor or a general contracting company using its own crews.

Material: Material costs are based on the national average contractor purchase price delivered to the job site. Delivered costs are assumed to be within a 20 mile radius of metropolitan areas. No sales tax is included in the material prices because of variations from state to state.

The prices are based on large quantities that would normally be purchased for projects costing $1,000,000 and up. Prices for small quantities must be adjusted accordingly. If more current costs for materials are available for the appropriate location, it is recommended that adjustments be made to the unit costs to reflect any cost difference.

Labor: The unit cost for labor of an item is calculated as shown previously. The daily labor cost of the designated crew is divided by the number of units that the crew will install in one day.

Equipment: The unit cost for equipment is calculated by dividing the daily bare equipment cost by the daily output.

The labor rates used to determine the bare installation costs are shown for 35 standard trades in Figure D.3, under the column "Base Rate Including Fringes." This rate includes a worker's actual hourly wage plus employer-paid benefits (health insurance, vacation, pension, etc.). The labor rates used are *national average* union rates based on trade union agreements (as of January 1 of the current year) from 30 major cities in the United States. As new wage agreements are negotiated within a calendar year, labor costs should be adjusted.

Total Bare Costs: This column simply represents the arithmetic sum of the bare material, labor, and equipment costs. This total is the average cost to the contractor for the particular item of construction, supplied and installed, or "in place." No overhead and/or profit is included.

Total Including Overhead and Profit

The prices in the "Total Including Overhead and Profit" column might also be called the "billing rate." These prices are, on the average, what the installing contractor would charge for the particular item of work. The term "installing contractor" can refer to either a subcontractor or a general contractor. In effect, this rate reflects the amount the subcontractor would charge to the general contractor or what the general contractor would charge the owner for work performed by the general contractor's own employees. The general contractor normally adds a percentage to subcontractor prices (commonly 10%) for supervision and management.

The "Total Including Overhead and Profit" costs are determined by adding the following two calculations:

1. Bare materials and equipment—increased by 10% for handling. See Figure D.2.
2. Bare daily crew labor costs—increased to include overhead and profit, then divided by the daily output.

In order to increase crew cost to include overhead and profit, labor and equipment costs are treated separately. 10% is added to the bare equipment cost for handling, management, etc. Labor costs are increased, depending upon trade by percentages for overhead and profit (as shown in Figure D.3). The resulting rates are listed in the right hand columns of Figure D.3. Note that the percentage increase for overhead and profit for the average skilled worker is 56.3% of the base rate. The following items are included in the increase for overhead and profit, as shown in Figure D.3.

Workers' Compensation and Employer's Liability:

Workers' Compensation and Employer's Liability Insurance rates vary from state to state and are tied into the construction trade safety records in that particular state. Rates also vary by trade according to the hazard involved. The proper authorities will most likely keep the contractor well informed of the rates and obligations.

State and Federal Unemployment Insurance: The employer's tax rate is adjusted by a merit-rating system according to the number of former employees applying for benefits. Contractors who find it possible to offer a maximum of steady employment can enjoy a reduction in the unemployment tax rate.

Employer-Paid Social Security (FICA): The tax rate is adjusted annually by the federal government. It is a percentage of an employee's salary up to a maximum annual contribution.

Builder's Risk and Public Liability: These insurance rates vary according to the trades involved and the state in which the work is done.

Overhead: The column listed as "Sub's Overhead" provides percentages to be added for office or operating overhead. This is the cost of doing business. The percentages are presented as national averages by trade as shown in Figure D.3. Note that the operating overhead costs are applied to *labor only* in *Means Building Construction Cost Data.*

Profit: This percentage is the fee added by the contractor to offer both a return on investment and an allowance to cover the risk involved in the type of construction being bid. The profit percentage may vary from 4% on large, straightforward projects to as much as 25% on smaller, high-risk jobs. Profit percentages are directly affected by economic conditions, the expected number of bidders, and the estimated risk involved in the project. For estimating purposes, *Means Building Construction Cost Data* assumes 10% as a reasonable average profit factor.

Cost Segregation Glossary

Accelerated Cost Recovery System (ACRS) A method of calculating depreciation of assets placed in service from 1980 to 1986 for federal income tax purposes. Replaced by the Modified Accelerated Cost Recovery System (MACRS) by the Tax Reform Act of 1986.

accelerated depreciation Depreciation at a faster rate than the straight-line method. Examples include the 200% or 150% declining balance methods.

accrual method The method of accounting whereby revenues and expenses are identified with specific periods of time, such as a month or a year, and are recorded as they are incurred without regard to the date of receipt or payment in cash.

actual cost records Contemporaneous construction and accounting records detailing actual costs from a constructed project, including invoices, contracts, subcontracts, change orders, and applications for payment. When used with a detailed engineering approach to cost segregation, these records provide the best results and the qualifying evidence.

addenda Any modifications, clarifications, or revisions to the contract documents, for any reason, issued by the architect during the bidding period of a construction project.

allocated cost The total cost of an asset's individual components, summarized, including indirect costs, and classified into one of four categories.

allowance (1) A stated requirement of construction contract documents whereby a specified sum of money is incorporated or allowed into the contract sum in order to sustain the cost of a stipulated material, assembly, piece of equipment, or other part of a construction contract. (2) A fixed lump sum to cover any items that have not yet been finalized and are not ready for inclusion in the bid set at the time the contract documents are ready to be issued.

alternates Additions or deletions to the base price of a construction contract. Alternates allow owners to see—during the bid phase of the project—how a change in materials, method of construction, or an addition or subtraction of work will affect the project's price.

Alternative Minimum Tax (AMT) A minimum amount of tax required to be paid by some taxpayers. The tax law gives special treatment to some kinds of income and allows special deductions and credits for some expenses. Taxpayers who benefit from the laws in these ways may have to pay at least a minimum amount of tax. In certain instances, AMT mandates slower depreciation methodologies, and its application may reduce some of the tax savings associated with cost segregation.

arc Any portion of the circumference of a circle.

architect's scale A draftsman's tool with proportionate, graduated spaces. May be flat, like a ruler, or three-sided. The three-sided scale has 10 separate scales: 1/8" and 1/4", 1" and 1/2", 3/4" and 3/8", 3/16" and 3/32", and 1-1/12" and 3".

architectural drawings Also called "core drawings," these show the layout of the building and its use of space. Architectural drawings convey the structure's aesthetic value and show the dimensions and placement of all key features.

architectural floor plan The most common plan view, which shows doors, windows, walls, and partitions.

area Method of measurement that refers to surfaces in two dimensions—length by width—expressed in square units.

as-built drawings Record drawings made during construction that document the final, completed project, including any subtle changes between the working drawing design and the way the structure is built.

asset An item of monetary value, which can include real, personal, or financial property. Usually expected to have some value in a future period.

basis (for depreciation) The basis for calculating depreciation of property is generally the cost to acquire it. If property is acquired by a method other than purchase, the basis is calculated differently (inheritance, gift, like-kind exchange, or self-constructed).

benchmark A known elevation on a project site for use as a reference point during construction. Benchmarks are established in reference to the datum, and are commonly noted on the site drawing with a physical description and elevation relative to the datum.

bid form A preprinted document used by the contractor/builder to submit an offer or "proposal" to perform a defined scope of work for a stipulated price. Bid forms are prepared by the architect or owner to allow comparision of price, since scope is the same for all bidders.

bidding period The time period between the date the drawings for a construction project are issued and day the bid is due.

billing rate The full value of a labor-hour, including all overhead, burden, and profit.

bonus depreciation Allowed in addition to regular depreciation, this 30% or 50% additional first year depreciation deduction is generally allowed for investment in qualifying new property acquired or constructed after September 10, 2001 and placed in service, in most cases, by December 31, 2004.

burdened labor cost A term describing a wage rate after it has been modified by adding all taxes, insurance, and overhead.

capitalized interest The interest incurred by a building owner on debt used to finance construction, and, in some cases, other interest that is required to be added to the basis of self-constructed assets.

cash method A method of accounting whereby revenues and expenses are recorded when the item of income or expense is actually paid or received in cash.

center line A light line of alternating long and short segments that indicates the center of a particular object. Frequently labeled with the letter "C" superimposed over the letter "L."

change order Written authorization provided to a contractor approving a change in scope from the original plans, specifications, or other contract documents. A change order can be a change in price, time, or both. It requires the agreement of the contractor, architect, and owner.

chord A straight line connecting two points on the circumference of a circle, without passing through its center.

circumference The measure of the perimeter of a circle.

civil drawings The group of different types of site drawings, such as utility and drainage, grading, site improvement, and landscaping plans. Civil drawings encompass all work other than the structure itself.

component depreciation The practice under prior law where taxpayers would separate a building's parts into its various components—doors, walls, roofs, floors, etc. Once these components were identified, taxpayers could depreciate them using different cost-recovery periods. The component method of depreciating buildings is no longer allowed. The building and its components are treated as one asset under current law.

construction documents The written specifications and drawings that describe the requirements of a construction project and can be used to build it.

contemporaneous documentation Actual project invoices, documents, plans, and specifications used to document the outlay of capital by a particular taxpayer for a particular construction project. Generally considered the best supporting documentation for a quality cost segregation study.

contour A line connecting points of equal elevation to show changes in the elevation of the lot, such as slopes, hills, valleys, and other variations in the surface.

contract documents All the written and graphic documents concerning execution of a particular construction contract. Contract documents most often consist of a set of plans, specifications, applications for payment or other types of invoicing, change orders or contract modifications, any relevant addenda or bulletins, and any additional relevant documents, such as geotechnical reports and documentation, or unique requirements of local authorities.

contract for construction An agreement between an owner and a contractor whereby the contractor agrees to construct the owner's project in accordance with the contract documents, within a specified amount of time and for consideration to be paid by the owner, as mutually agreed.

convention Intended to simplify depreciation calculations, averaging conventions assume a placed-in-service date that is uniform for all assets within the same class of property. The half-year convention assumes assets are placed in service at the mid-point of the tax year.

core drawings The architectural set of drawings. All other disciplines, such as mechanical, electrical, and structural drawings, are designed around the core drawings.

cost segregation A process of identifying assets within a larger construction project and using cost estimating techniques that allocate costs to individual items of property, which can then be classified for depreciation purposes (e.g., land improvements, buildings, equipment, furniture and fixtures, etc.). The underlying incentive for building owners is the significant acceleration of tax benefits (depreciation deductions) resulting in improved current after-tax cash flow.

crew tasks Tasks done by crews that combine the performance of different individuals, versus a single individual. Each crew member performs a specific function to complete the overall task.

CSI MasterFormat System developed by the Construction Specifications Institute to organize the technical specifications for construction products and services. CSI has allocated a five-digit code and topic descriptions to all components of the specifications. MasterFormat groups the technical specifications information into 16 divisions, each of which is a grouping of similar or related work numerically organized into subsections.

datum The known point of reference from which elevation is measured, such as sea level.

demolition safe harbor When a portion of the building is demolished, if 75% of the building's internal and external structure is left untouched, then the modification to the building is not treated as a demolition for tax purposes.

depreciation An annual deduction allowed to recover the cost or other basis of property having a useful life beyond the current year in which it is used in a trade or business or for the production of income. Usually represents the decline in property value due to age, general wear and tear, obsolescence, and so forth, over the life of the asset.

design development The process of reviewing and modifying construction drawings as a result of owner input, coordination with other design disciplines, building code compliance, and general fine-tuning.

detail An enlargement of part of a plan, elevation, or drawing that provides greater clarification and understanding.

detailed engineering cost estimate approach A methodology for determining the costs of assets within a construction project. Using contemporaneous construction and accounting records with estimated cost information. Also referred to as *detailed engineering actual cost approach.*

dimension line A light, fine line with arrowheads or "tic" marks at each end, used to show the measurements of the main object lines.

direct overhead costs Project overhead costs that are directly related to one individual project. These may include temporary facilities, trailer rental, superintendence, telephone usage, staging or lifts, and electrical power.

dwelling units Portions of a buiding (apartments) used to provide living accommodations in a building or structure on a non-transient basis. Length of stay is generally more than 30 days.

electrical drawings Graphic representations of the electrical requirements of a project, including power distribution, lighting, and low-voltage specialty wiring, such as for fire alarms, telephone/data, and technology wiring.

elevation Graphic format that provides a pictorial view of the walls of a structure, similar to a photograph of a wall taken perpendicular to both the vertical and horizontal planes.

engineer's scale A draftsman's tool with proportionate, graduated spaces used to indicate distances on site plans when these distances are greater than would be encountered in the building plans. The scale is based on number of feet in real life per inch on the drawings and has six scales: 10, 20, 30, 40, 50, and 60.

estimate replication *See detailed engineering cost estimate approach.*

extended cost The resultant cost once a unit price has been multiplied by the quantity, also called the *total cost* for the item.

extension line A light line that extends from the edge or end of the main object line. Used together with dimension lines to help determine the dimension of a particular feature.

fixed project overhead costs Direct overhead costs that are not affected by time.

general liability insurance Insurance that protects a project and any adjacent properties from damage or loss as a result of an accident during the construction process.

grading plan A separate plan in the contract documents that indicates topography. Grading plans are used for projects that require the topography to be shown separately for clarity. Grading plans also indicate how surface precipitation will be channeled away from the structure.

indirect costs Costs of a construction project that may not be identified to a specific item of property. These are accumulated and allocated in total to the separately identified assets at the end of the cost segregation study. May include costs that are not in the construction contract, such as design consultants, capitalized interest, and other items capitalized under the uniform capitalization rules.

indirect overhead costs Also referred to as *main office overhead*, any costs of a general nature that are not unique to a specific project, but are related to the operation of the contractor's main office and its staff—costs associated with being in business.

Internal Revenue Code (IRC) Common term used to describe the collection of federal laws enacted by Congress that impose taxes. The most common of these is the tax based on the taxable income of individuals, estates, trusts, and corporations. Also encompasses the rules of administration and procedures to be used by the Internal Revenue Service in the determination of any tax.

land Unimproved real property, whose cost sometimes includes tree removal and site construction costs incurred to achieve a level grade or to make the property ready for its intended use. Land is one of the four categories into which assets can be classified, along with the building, land improvements, and personal property.

land improvements Constructed assets permanently affixed to land usually used in conjunction with buildings. Common examples include sidewalks, flagpoles, and parking lots. One of the four categories into which assets can be classified.

locus A small map included on a site plan that shows the general location of a property in respect to local highways, roads, and adjacent pieces of property.

main object line A thick, heavy, unbroken line on construction drawings that defines the outline of the structure or object. Used for the main outlines of walls, floors, elevations, details, or sections.

mechanical drawings Graphic depiction of a structure's physical systems, such as plumbing, fire suppression/protection, and HVAC (heating, ventilating, and air conditioning) systems.

Modified Accelerated Cost Recovery System (MACRS) The currently required method for determining the calculation of depreciation for tax purposes. MACRS comprises two subsystems: the General Depreciation System (GDS) and the Alternative Depreciation System (ADS).

Net Present Value (NPV) The discounted value of future cash flows (tax savings) in terms of current dollars. Current tax savings (deferral of taxes) are achieved by reclassifying costs to a tangible property classification that would otherwise be classified as 39-year or 27.5-year property.

non-production costs Overhead costs, or the cost of supporting actual production activities.

non-residential rental real estate Buildings and structural components of buildings that are not residential rental real estate. Non-residential rental real estate is depreciated using the straight-line method over 39 years.

paving/curbing layout drawing Graphic layout of parking lots and driveways. These drawings show the various types of bituminous concrete, brick paving and curbing, and the limits of each, and are helpful for calculating areas and measurements.

personal property (tangible) One of the four categories into which assets can be classified. Personal property is all tangible property other than land and permanent improvements to land, such as buildings and their structural components. Examples include transportation and office equipment, refrigerators, and display racks. Taxpayers normally can depreciate personal property using a 5- or 7-year recovery period and the double-declining balance method.

plan view The most common graphic format, presented as if looking down on the space.

production costs Part of the actual, physical project, including materials, labor, tools, equipment, and subcontractor costs for each task or activity incorporated in the final structure.

productivity The rate at which work is performed by an individual or crew per unit of time—usually per day.

qualified leasehold improvement property Improvements to tenant-occupied space in a building that is non-residential real property, made by a landlord or tenant under a lease. Under certain circumstances may qualify for the additional first year depreciation deductions of 30% or 50%.

quantity takeoff The process in which detailed lists are compiled of all the tasks, labor, and materials necessary to construct a project, based on drawings and specifications. It is a basic function of estimating.

real property Land and improvements to land, such as buildings. *See also Section 1250 property.*

requisition To apply for partial progress payments based on work completed at the end of the pay period (usually one month). Abbreviated term for application for payment.

residential rental real estate Any building or structure from which 80% or more of the rental revenues come from rentals of dwelling units.

residual approach A methodology for determining the costs of an asset within a construction project. An abbreviated method that determines the cost of assets with a short life span (5 and 7 years) and the residual is allocated to the building classifications.

rule of thumb approach A methodology for determining the costs of assets within a construction project. Little or no documentation is relied on, as the practitioner uses personal experience and knowledge of the particular industry and approximates Section 1245 property as a percentage of construction costs.

sampling approach A methodology for determining the costs of assets within a construction project. Uses a model to analyze identical types of construction and is most popular for redundant type projects, such as retail stores. Relies heavily on the validity of the sample or model results. Individual properties may be affected by geography, local building codes, and labor costs.

schedule of values Itemization of work tasks on a line-by-line basis, which lists the dollar value of each task at 100% complete.

Section 1250 property any tangible property that is not Section 1245 property. The definition of Section 1245 property excludes land and improvements to land, such as buildings and other permanent structures, including items that are structural components of such buildings. *See also: non-residential real property and residential rental property.*

Section 1245 property All tangible property other than land and permanent improvements to land, such as buildings and their structural components. Includes assets that can be shown to be accessory to the operation of a trade or business.

setback dimensions The distance to property lines, shown on a site plan in feet and hundredths of a foot (versus feet and inches on architectural drawings).

shading The effect architects and engineers use to convey information on a drawing in a subtle manner by changing the intensity of a particular feature.

site drawings Also called *civil drawings*, these illustrate a structure's relationship to the property, including various engineering improvements to the site, such as the sanitary system, utilities, paving walks, curbing, and so forth.

site improvements drawings Graphic representation of ordinary site improvements, such as fencing, patios, walks, flagpoles, and the like.

site plan Included on commercial projects, a site plan illustrates the relationship of the proposed structure to the building's lot, as well as the various site improvements needed to accommodate the new building.

sole justification test Found in regulations Section 1.48-1, used to classify machinery whose purpose is temperature and humidity control related to the operation of other machinery or to a process, and therefore may not be considered a "structural component" of a building.

specialty drawings Graphic representations of the unique requirements of various spaces' special uses (such as kitchens, libraries, retail spaces, and home theater systems). Specialty drawings define the coordination among other building systems, most commonly the mechanical and electrical systems.

specification–performance An alternate method of specifying products and materials, based less on particular models and more on the ability to satisfy a design requirement or perform a specific function. In lieu of specifying a particular product by name, the architect opens competition to all parties that can perform the specific functions required to complete the design.

specifications Part of the contract documents, used along with the working set of drawings, that define in detail the processes and materials for the project. The specs perform a variety of functions, including: serving as the legal basis for the contract for construction, defining the quality or grade of materials to be used in the project, defining the acceptable workmanship or providing standards to judge workmanship, providing guidelines for resolving disputes between parties to the contract, providing a basis for accurately estimating cost, and complementing the graphic portion of the project, the drawings.

specifications–descriptive Written instructions or details for assembling various components to comprise a system or assembly. Most often, descriptive specifying is used for generic products such as mortar, lumber, or concrete.

specifications–technical Definitions of the scope, products, and execution of the work. Technical specifications provide the necessary information to accurately price assets or the entire structure. Architects and engineers rely on them to establish the quality of the materials and workmanship to be incorporated in the project.

state investment tax credits Credits allowed to taxpayers for constructing projects for such purposes as manufacturing or rehabilitating historic buildings. Usually based on a percentage of qualifying costs. The determination of costs eligible may require an analysis of proper depreciable lives and proper allocation for tax credit purposes. Another potential use for cost segregation studies.

straight-line method of depreciation A depreciation method that spreads the depreciation evenly over the life of the property. The rate is calculated by dividing 1 by the life of the asset. The annual depreciation deduction for a 10-year asset would be 10% (1/10).

structural components For tax depreciation purposes, any property that relates to the operation or maintenance of a building, including building parts (walls, floors, and ceilings); permanent coverings (paneling, windows, and doors); central air conditioning or heating system (motors, pipes, and ducts); plumbing and fixtures (sinks and bathtubs); electrical wiring and lighting fixtures; stairs and elevators; and sprinkler systems.

structural drawings Graphic representations of the members, assemblies, and systems that will transmit live and dead loads of the structure to the earth.

survey or letter approach A methodology for determining the costs of assets within a construction project. Requests are made to the contractor/builder for specific identified costs. These are provided and used in conjunction with one of the other methods, typically one of the detailed engineering approaches or the residual estimation approach.

surveys A way to acquire supporting information to produce a full set of asset costs. These inquiries are sent to contractors, subcontractors, and vendors to request cost information on specific assets or categories of work performed by those firms.

tangible personal property All tangible property, except land and inherently permanent improvements to land, such as buildings and their structural components. *See also Section 1245 property and personal property.*

Tax Court The U.S. federal court that hears tax-related cases between taxpayers and the Internal Revenue Service.

tax deferral Methodology that results in a change in the period in which income tax is ultimately paid. Using future deductions in the current period results in lower current year taxable income and, accordingly, a lower tax. In later years, the tax will be higher, since the same deduction will not be available in that future period to reduce taxable income.

taxable income Gross income minus allowable deductions.

time-sensitive costs Costs defined by the duration for which they are needed. The longer the particular item (such as a dumpster) is on the job site, the higher the cost.

uniform capitalization rules Requires the capitalization of direct and certain indirect expenses of building construction projects. Includes interest, project supervision, and other indirect costs that benefit the construction project. Applies to most real estate construction. *(See IRC Section 263A and the related regulations.)*

unit price Current and accurate cost of materials, equipment, and labor used to develop a unit price estimate.

unit price estimate An estimate that assigns an accurate quantity and price to each item of work or asset. Performed by breaking a project down into tasks, quantifying those tasks, and then applying a price based on the number of units for each task.

utility drawings Graphic depictions of locations of water, gas, sanitary sewer, and electric utilities that will service the building.

Worker's Compensation Insurance to protect employees, provided by employers in the event of injury, disablement, or death occurring in the workplace.

wage rate Hourly rate of pay on which the employee's paycheck is calculated, agreed upon between the employer and employee.

waste The difference between the quantity that must be included in an estimate and the quantity that is actually to be installed, added to the quantity as a percentage. Applicable to materials, waste should not be applied to labor or confused with productivity.

working drawings The final, completed design. A code-compliant representation of the project, ready for construction.

Resources

Recommended Reading

Reference Books from RSMeans

Electrical Estimating Methods, Third Edition
Estimating Building Costs for the Residential & Light Commercial Contractor by Wayne DelPico
Mechanical Estimating Methods, Third Edition
Means Estimating Handbook, Second Edition
Means Illustrated Construction Dictionary, Unabridged, Third Edition
Plan Reading & Material Takeoff by Wayne DelPico
Plumbing Estimating Methods, Third Edition
Unit Price Estimating Methods, Third Edition

Cost Data from RSMeans

Building Construction Cost Data
Concrete & Masonry Cost Data
Electrical Cost Data
Facilities Construction Cost Data
Mechanical Cost Data
Open Shop Building Construction Cost Data
Plumbing Cost Data
Site Work & Landscape Cost Data
Square Foot Cost Data

Software

Means *CostWorks*
Web site: http://www.rsmeans.com/supplement/67345.asp

Tax Resources

Internal Revenue Service
http://www.irs.gov
Publication 946
"Form 4562"
Cost Segregation Audit Techniques Guide

Authored by a team of IRS examiners, engineers, and attorneys, this comprehensive document provides guidance for taxpayers and practitioners in preparing cost segregation studies. Available online at: http://www.irs.gov/businesses/article/0,,id=129356,00.html

United States Tax Court
400 Second Street, NW
Washington, DC 20217
202-606-8754
http://www.ustaxcourt.gov

Publications

Federal Income Taxation of Real Estate
Warren Gorham & Lamont
Available electronically or in print, comprehensive analysis of real estate tax laws and regulations, as well as guidance on incorporating the latest tax amendments.

Practitioners Publishing Company
Publishes several construction and real estate resource guides designed for accounting professionals.
http://www.ppcnet.com

U.S. Master Depreciation Guide
CCH Incorporated
Annually updated guidance on understanding and applying depreciation rules to fixed assets.

Index

C

J

K

L

M

N

O

P

Q

R

S

T

www.ingramcontent.com/pod-product-compliance
Lightning Source LLC
LaVergne TN
LVHW062148200926
848484LV00007B/1881